Sebastian Höppner
Clock Generator Circuits for Low-Power
Heterogeneous Multiprocessor Systems-on-Chip

TUD*press*

Sebastian Höppner

Clock Generator Circuits for Low-Power Heterogeneous Multiprocessor Systems-on-Chip

TUD*press*
2013

Die vorliegende Arbeit wurde am 28. Februar 2013 an der Fakultät Elektrotechnik und Informationstechnik der Technischen Universität Dresden als Dissertation eingereicht und am 25. Juli 2013 erfolgreich verteidigt.

Vorsitzender: Prof. Dr. rer. nat. Johann W. Bartha

Gutachter:
Prof. Dr.-Ing. habil. René Schüffny
Prof. Dr.-Ing. Ulrich Rückert

Bibliografische Information der Deutschen Nationalbibliothek
Die Deutsche Nationalbibliothek verzeichnet diese Publikation in der Deutschen Nationalbibliografie; detaillierte bibliografische Daten sind im Internet über http://dnb.d-nb.de abrufbar.

Bibliographic information published by the Deutsche Nationalbibliothek
The Deutsche Nationalbibliothek lists this publication in the Deutsche Nationalbibliografie; detailed bibliographic data are available in the Internet at
http://dnb.d-nb.de.

ISBN 978-3-944331-20-1

Verlag der Wissenschaften GmbH
Bergstr. 70 | D-01069 Dresden
Tel.: 0351/47 96 97 20 | Fax: 0351/47 96 08 19
http://www.tudpress.de

Gesetzt vom Autor.
Printed in Germany.

Technische Universität Dresden

Clock Generator Circuits for Low-Power Heterogeneous Multiprocessor Systems-on-Chip

Sebastian Höppner

von der Fakultät Elektrotechnik und Informationstechnik der
Technischen Universität Dresden

zur Erlangung des akademischen Grades eines

Doktoringenieurs

(Dr.-Ing.)

genehmigte Dissertation

Vorsitzender:	Prof. Dr. rer. nat. Johann W. Bartha		
Gutachter:	Prof. Dr.-Ing. habil. René Schüffny	Tag der Einreichung	28.02.2013
	Prof. Dr.-Ing. Ulrich Rückert	Tag der Verteidigung	25.07.2013

Für Anja und Marek

Danksagung

Ich möchte mich bei meinem Betreuer Herrn Prof. René Schüffny für die Möglichkeit bedanken, dieses interessante Thema bearbeiten zu können. Seine fachliche Betreuung sowie das entgegengebrachte Vertrauen selbstbestimmt eigene Strategien zur Realisierung dieser komplexen Inhalte verfolgen zu können, haben maßgeblich beigetragen diese Arbeit erfolgreich zum Abschluss zu bringen.
Frau Prof. Renate Merker gilt mein Dank für die kritischen und wertvollen Diskussionen zum Verfassen dieser Arbeit.
Insbesondere bedanke ich mich bei meinen Kollegen von der Stiftungsprofessur für Hochparallele VLSI-Systeme und Neuromikroelektronik sowie der RacyICs GmbH für das stets angenehme und freundschaftliche Arbeitsklima. Besonders erwähnen möchte ich hier Holger Eisenreich für die Konzipierung und Koordinierung der komplexen Systems-on-Chip, welche maßgeblich die Inhalte dieser Arbeit motiviert haben. Ich danke Georg Ellguth für die exzellente und ausdauernde Arbeit an der physischen Realisierung der Testchips und Dennis Walter für die wesentlichen Beiträge zur Realisierung der Standardzell-Bibliotheken. Nicht zuletzt gilt mein Dank Dr. Stephan Henker und Stefan Hänzsche für die vielen tiefgehenden Fachdiskussionen, welche meine Tätigkeit stets inspiriert haben. Ohne die Unterstützung durch dieses starke Team von Ingenieuren wäre diese Arbeit nicht realisierbar gewesen.
Meinen Eltern Kristina und Wolfgang danke ich für die jahrelange Unterstützung meiner Ausbildung sowie auch dafür, dass sie immer ein offenes Ohr für mich hatten. Sie haben mir stets das Selbstvertrauen gegeben, dass es sich lohnt, zielstrebig meine Ziele umzusetzen.
Ein besonderer Dank gilt meiner lieben Anja und unserem kleinen Sohn Marek für das entgegengebrachte Verständnis und die Geduld mit mir, wenn ich oftmals viel Zeit im Büro verbringen musste. Ihr seid mir ein starker Rückhalt und habt mich doch stets daran erinnert, dass Arbeit nicht alles im Leben ist.

Abstract

In this work concepts and circuits for local clock generation in low-power heterogeneous multiprocessor systems-on-chip (MPSoCs) are researched and developed. The targeted systems feature a globally asynchronous locally synchronous (GALS) clocking architecture and advanced power management functionality, as for example fine-grained ultra-fast dynamic voltage and frequency scaling (DVFS). To enable this functionality compact clock generators with low chip area, low power consumption, wide output frequency range and the capability for ultra-fast frequency changes are required. They are to be instantiated individually per core. For this purpose compact all digital phase-locked loop (ADPLL) frequency synthesizers are developed. The bang-bang ADPLL architecture is analyzed using a numerical system model and optimized for low jitter accumulation. A 65nm CMOS ADPLL is implemented, featuring a novel active current bias circuit which compensates the supply voltage and temperature sensitivity of the digitally controlled oscillator (DCO) for reduced digital tuning effort. Additionally, a 28nm ADPLL with a new ultra-fast lock-in scheme based on single-shot phase synchronization is proposed. The core clock is generated by an open-loop method using phase-switching between multi-phase DCO clocks at a fixed frequency. This allows instantaneous core frequency changes for ultra-fast DVFS without re-locking the closed loop ADPLL. The sensitivity of the open-loop clock generator with respect to phase mismatch is analyzed analytically and a compensation technique by cross-coupled inverter buffers is proposed.

The clock generators show small area (0.0097mm^2 (65nm), 0.00234mm^2 (28nm)), low power consumption (2.7mW (65nm), 0.64mW (28nm)) and they provide core clock frequencies from 83MHz to 666MHz which can be changed instantaneously. The jitter performance is compliant to DDR2/DDR3 memory interface specifications. Additionally, high-speed clocks for novel serial on-chip data transceivers are generated. The ADPLL circuits have been verified successfully by 3 testchip implementations. They enable efficient realization of future low-power MPSoCs with advanced power management functionality in deep-submicron CMOS technologies.

Kurzfassung

In dieser Arbeit werden Konzepte und Schaltungen zur lokalen Takterzeugung in heterogenen Multiprozessorsystemen (MPSoCs) mit geringer Verlustleistung erforscht und entwickelt. Diese Systeme besitzen eine global-asynchrone lokal-synchrone Architektur sowie Funktionalität zum Power Management, wie z.B. das feingranulare, schnelle Skalieren von Spannung und Taktfrequenz (DVFS). Um diese Funktionalität zu realisieren werden kompakte Taktgeneratoren benötigt, welche eine kleine Chipfläche einnehmen, wenig Verlustleitung aufnehmen, einen weiten Bereich an Ausgangsfrequenzen erzeugen und diese sehr schnell ändern können. Sie sollen individuell pro Prozessorkern integriert werden. Dazu werden kompakte volldigitale Phasenregelkreise (ADPLLs) entwickelt, wobei eine bang-bang ADPLL Architektur numerisch modelliert und für kleine Jitterakkumulation optimiert wird. Es wird eine 65nm CMOS ADPLL implementiert, welche eine neuartige Kompensationsschlatung für den digital gesteuerten Oszillator (DCO) zur Verringerung der Sensitivität bezüglich Versorgungsspannung und Temperatur beinhaltet. Zusätzlich wird eine 28nm CMOS ADPLL mit einer neuen Technik zum schnellen Einschwingen unter Nutzung eines Phasensynchronisierers realisiert. Der Prozessortakt wird durch ein neuartiges Phasenmultiplex- und Frequenzteilerverfahren erzeugt, welches es ermöglicht die Taktfrequenz sofort zu ändern um schnelles DVFS zu realisieren. Die Sensitivität dieses Frequenzgenerators bezüglich Phasen-Mismatch wird theoretisch analysiert und durch Verwendung von kreuzgekoppelten Taktverstärkern kompensiert.

Die hier entwickelten Taktgeneratoren haben eine kleine Chipfläche (0.0097mm^2 (65nm), 0.00234mm^2 (28nm)) und Leistungsaufnahme (2.7mW (65nm), 0.64mW (28nm)). Sie stellen Frequenzen von 83MHz bis 666MHz bereit, welche sofort geändert werden können. Die Schaltungen erfüllen die Jitterspezifikationen von DDR2/DDR3 Speicherinterfaces. Zusätzliche können schnelle Takte für neuartige serielle on-Chip Verbindungen erzeugt werden. Die ADPLL Schaltungen wurden erfolgreich in 3 Testchips erprobt. Sie ermöglichen die effiziente Realisierung von zukünftigen MPSoCs mit Power Management in modernsten CMOS Technologien.

Contents

List of Figures

List of Tables

Acronyms

ADPLL all-digital phase-locked loop

ADC analog-to-digital converter

AVFS adaptive voltage and frequency scaling

BBADPLL bang-bang ADPLL

BBPFD bang-bang PFD

BER bit-error-rate

CCO current-controlled oscillator

CMOS complementary metal oxide semiconductor

CP charge pump

CPU central processing unit

DAC digital-to-analog converter

DCO digitally controlled oscillator

DCOP DC operating point

DDR double data rate

DFM design for manufacturability

DFT design for test

DFY design for yield

DLL delay-locked loop

DNL differential nonlinearity

DRC design rule check

DSM delta-sigma modulator

DSP digital signal processor

DVFS dynamic voltage and frequency scaling

EDA electronic design automation

FFT fast Fourier transform

FIFO first in, first out

FPGA field programmable gate array

FSM finite state machine

GALS globally asynchronous locally synchronous

GPU graphics processing unit

HPM hardware performance monitor

HVT high threshold voltage

ILM interface logic model

IP intellectual property

JTAG Joint Test Action Group

LOP linearized operating point

LSB least significant bit

LVDS low voltage differential signaling

LVS layout versus schematic

LVT low threshold voltage

MIMO multiple input multiple output

MPSoC multiprocessor system-on-chip

MSB most significant bit

MOS metal oxide semiconductor

NMOS n-channel metal oxide semiconductor transistor

NoC network-on-chip

OTA operational transconductance amplifier

PCB printed circuit board

PDK process design kit

PFD phase frequency detector

PL performance level

PLL phase-locked loop

PMIC power management IC

PMOS p-channel metal oxide semiconductor transistor

PMU power management unit

PVT process, voltage and temperature

RF radio frequency

RISC reduced instruction set computing

RTL register transfer level

SLVT super low threshold voltage

SoC system-on-chip

SRAM static random access memory

STA static timing analysis

SDR single data rate

TDC time-to-digital converter

TSV through-silicon via

VCO voltage controlled oscillator

VDSP vector DSP

List of Symbols

symbol	unit	description
$\mathbf{A}$		current compensation coefficient matrix
a	[1]	logic toggle rate
a_{comp}	[A/V]	current compensation coefficient, voltage related
$\mathbf{a}$		current compensation coefficient vector
b_{comp}	[A/°C]	current compensation coefficient, temperature related
C	$[F]$	capacitance
C'	$[F/\mathrm{m}]$	capacitance per width of MOS transistor
$C_{\mathrm{buf,in}}$	$[F]$	buffer input capacitance
C_{int}	$[F]$	internal capacitance
C_{muxnode}	$[F]$	multiplexer tristate node capacitance
C_{on}	$[F]$	on capacitance
C_{off}	$[F]$	off capacitance
C_{par}	$[F]$	parasitic capacitance
C_{L}	$[F]$	load capacitance
$C_{\mathrm{tri,out}}$	$[F]$	tristate driver output capacitance
C_{wire}	$[F]$	wire capacitance
c	[1]	DCO tuning signal , normalized
c_0	[1]	DCO tuning signal , nominal value
c_{ao}	[1]	static always-on DCO tuning signal
c_{coarse}	[1]	DCO coarse tuning signal
c_{comp}	[A]	current compensation coefficient
c_{DSM}	[1]	DSM modulated DCO tuning signal
$c_{\mathrm{DIV\_SEL}}$	[1]	division ratio select signal
c_{fine}	[1]	DCO fine tuning signal
c_{ftgain}	[1]	DCO fine tuning gain select signal
D	[1]	number of delay clock cycles
d	[1]	DSM output value
E	$[J]$	energy

E_{task}	$[J]$	energy of processor task
F		function, general
FOM_J	[1]	figure of merit for ADPLL period jitter performance
f	[Hz]	frequency
f_{core}	[Hz]	core clock frequency
f_{DCO}	[Hz]	DCO clock frequency
$f_{\text{flicker,corner}}$	[Hz]	flicker noise corner frequency
f_{NoC}	[Hz]	NoC clock frequency
G_0	[S]	nominal conductance
$G_{I,\text{ref}}$	[A]	reference current function
G_{tune}	[S]	tuning conductance
g_{m}	[S]	transconductance
H	[1]	transfer function
I	[A]	current, general
I_{D}	[A]	drain current
I_{D0}	[A]	on-current of MOS transistor
I_{DCO}	[A]	DCO current
I_{ref}	[A]	reference current
$I_{\text{ref,0}}$	[A]	reference current, constant part
$I_{\text{ref,comp}}$	[A]	compensated reference current
$I_{\text{ref,1,ptk}}$	[A]	reference current, positive temperature sensitivity
$I_{\text{ref,2,pvk}}$	[A]	reference current,positive voltage sensitivity
I_{tune}	[A]	DCO tuning current
I_{leak}	[A]	leakage current
j_{cc}	[s]	cycle-to-cycle jitter
j_T	[s]	period jitter
$j_{T,\text{pp}}$	[s]	period jitter peak to peak value
$j_{t,\text{abs}}$	[s]	absolute jitter
$j_{T,\text{acc},n}$	[s]	accumulated jitter over n cycles
K_{t}	[s]	DCO tuning gain
K	[1]	MOS transistor width ratio
K_{FO}	[1]	MOS transistor width ratio of fanout load
KP	[A/V^2]	MOS transistor transconductance parameter
k	[1]	index variable
$\mathbf{k}$	[1]	current compensation weighting vector
k_0, k_1, k_1	[1]	current compensation weighting factors
k_{GS}	[1]	LOP model parameter

L	[m]	transistor channel length
M	[1]	number of phases of a multi-phase DCO output signal
N	[1]	division ratio of PLL loop divider
$N_{\mathrm{drv,on}}$	[1]	number of activated tristate inverters
N_{NoC}	[1]	frequency multiplication ratio of NoC clock
N_{olclkg}	[1]	division ratio of open-loop clock generator
N_{sync}	[1]	open-loop clock generator synchronous division ratio
N_{tap}	[1]	number of DLL delay taps
n	[1]	number, general
n_{count}	[1]	counter value
n_{f}	[1]	number of filter cycles
n_{sw}	[1]	number of phase switchings
n_{step}	[1]	number of phases to be switched
n_{task}	[1]	number of clock cycles per task
P	[W]	power consumption
P_{ctrl}	[W]	controller power consumption
P_{DCO}	[W]	DCO power consumption
P_{olclkg}	[W]	open-loop clock generator power consumption
$P_{\mathrm{phasesync}}$	[W]	phase synchronizer power consumption
R	[Ω]	resistance
R_{on}	[Ω]	on-resistance of MOS transistor
R_{tune}	[Ω]	tuning resistance
S	[1]	sensitivity
S_{core}	[1]	mismatch jitter sensitivity of NoC clock
S_{NoC}	[1]	mismatch jitter sensitivity of NoC clock
S_{ϕ}	[dB/Hz]	phase noise power spectral density
T	[s]	period, general
T_{core}	[s]	period of core clock
T_{CLK}	[s]	clock period
T_{DCO}	[s]	period of DCO
T_{DIV}	[s]	period of divider clock
T_{NoC}	[s]	period of NoC clock
T_0	[s]	period, nominal value
T_{meas}	[s]	period of the measurement oscillator
T_{offset}	[s]	period, offset value
T_{ref}	[s]	period of reference clock
T_{step}	[s]	period step

t	[s]	time, general
$t_{\mathrm{clk,H}}$	[s]	clock high pulse time
$t_{\mathrm{clk,L}}$	[s]	clock low pulse time
t_{d}	[s]	delay time
$t_{\mathrm{d,0}}$	[s]	delay time, nominal value
$t_{\mathrm{d,intrinsic}}$	[s]	intrinsic delay time of CMOS stage
t_{dead}	[s]	dead zone time
t_{div}	[s]	absolute time of a frequency divider edge event
t_{err}	[s]	timing error
t_{H}	[s]	hold time
t_{jitter}	[s]	jitter time
t_{lock}	[s]	lock-in time
t_{offset}	[s]	offset time
Δt_{PFD}	[s]	PFD input timing difference
t_{ref}	[s]	absolute time of a reference clock edge event
t_{S}	[s]	setup time
t_{skew}	[s]	clock skew time
t_{step}	[s]	phase switching timing step
t_{task}	[s]	core task execution time
t_{tt}	[s]	transition time
V	[V]	voltage, general
V_{CM}	[V]	common-mode voltage
V_{DS}	[V]	drain source voltage
V_{DD}	[V]	supply voltage
$V_{\mathrm{DD,tune}}$	[V]	DCO tuning supply voltage
V_{GS}	[V]	gate source voltage
V_{in}	[V]	input voltage
V_{out}	[V]	output voltage
V_{SSnoc}	[V]	NoC link auxiliary supply level
V_{SB}	[V]	source bulk voltage
V_{th}	[V]	threshold voltage
W	[m]	transistor channel width
W_{P}	[m]	PMOS transistor channel width
W_{N}	[m]	NMOS transistor channel width
y	[1	count value
y_{ref}	[1	reference count value
α	[1]	loop filter integral coefficient

α	[1]	exponent in MOS transistor power-law model
β	[1]	loop filter proportional coefficient
β	[A/V^2]	MOS transistor transconductance parameter
γ	[1]	constant
ϵ_{step}	[1]	relative delay variation
ϕ	[1]	integrator register value
ϕ_{start}	[1]	initial integrator register value
ϕ_{dsm}	[1]	DSM integrator value
σ_{jitter}	[s]	jitter time standard deviation
$\sigma_{t,\text{abs}}$	[s]	standard deviation of the absolute jitter
$\sigma_{\Delta t}$, PFD	[s]	standard deviation of the PFD input time difference
σ_{cc}	[s]	standard deviation of the cycle-to-cycle jitter
$\sigma_{T,\text{acc},n}$	[s]	standard deviation of the accumulated jitter over n cycles
σ_T	[s]	standard deviation of the period jitter
$\sigma_{T,\text{core}}$	[s]	standard deviation of the core clock period jitter
$\sigma_{T,\text{DCO}}$	[s]	standard deviation of the DCO period jitter
$\sigma_{T,\text{ref}}$	[s]	standard deviation of the reference clock period jitter
$\sigma_{t_{\text{PFD}}}$	[1]	standard deviation of the PFD input timing difference
$\sigma_{t,\text{step}}$	[1]	standard deviation of the phase switching timing step
$\sigma_{t,\text{step},n}$	[1]	standard deviation of the phase switching timing step over n phases
τ	[s]	RC time constant
θ	[°C]	temperature

1 Introduction

Multiprocessor Systems on Chip (MPSoCs) are integrated circuits which contain multiple processing cores. They are key components of modern electronic systems and widely used for applications in mobile communications [Ram07, LWB+08], automotive electronics [HHN+10] or high-performance computing [RBB+11, MKO+12]. The integration of high compute power and various functionality within a single silicon die can significantly reduce its mechanical footprint, costs and energy consumption. Especially costly off-chip data transmission can be reduced significantly if more of the overall system components are integrated into one chip.

The trend towards multi-core processor chips has been heavily driven by desktop computing and high-performance computing, where an increased number of cores can directly provide higher computational performance. In these application scenarios low power consumption is not the primary system optimization target. Typically these systems are homogeneous MPSoCs, containing a number of equal cores [VHR+08, RBB+11, MKO+12]. In contrast to that, heterogeneous MPSoCs contain various cores with specialized functionality. A current trend is the integration of central processing units (CPUs) and graphics processing units (GPUs) into a system-on-chip (SoC) [DGJ+12] for mobile and desktop computing. A significantly higher degree of core diversity can be found in MPSoCs for mobile communication systems [Ram07], where it is essential to support a wide range of communication standards, while being restricted to low power consumption when these MPSoCs are used in mobile devices. Typically communication MPSoCs include general purpose processors (e.g. reduced instruction set computing (RISC) cores), digital signal processors (DSPs) [LWB+08] and dedicated hardware accelerators for communication algorithms (e.g. fast Fourier transform (FFT) [JSC+12], multiple input multiple output (MIMO) detection [WKA+12]). The powerful signal processing capability of modern MPSoCs requires high bandwidth I/O interfaces, e.g. to field programmable gate arrays (FPGAs) [SEH+12, SSP+11] or external memory chips via DDR2 or DDR3 interfaces [JED09], [JED10]) with low power consumption. Fig. 1.1 shows an example schematic of a heterogeneous MPSoC.

Clock signals are required in nearly all of todays digital circuits to define their

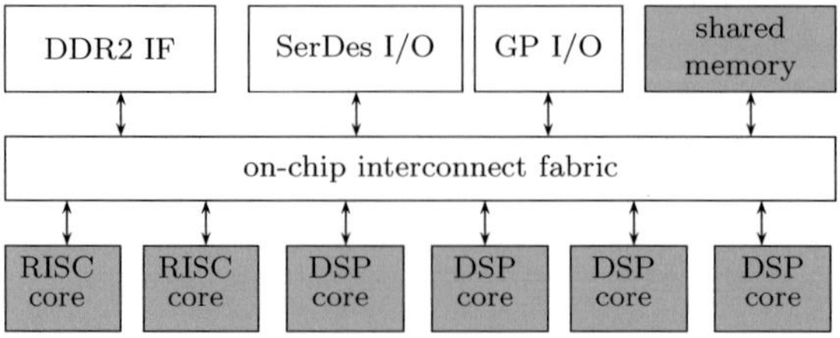

Figure 1.1: Heterogeneous MPSoC example block diagram

internal timing. Depending on the application there are various constraints for these clocks ranging from defined frequencies to good signal quality, as for example low jitter. Modern MPSoCs typically contain tens to hundreds of different clock domains. These clock signals have to be generated on-chip to minimize the number of global I/O pins and external components. Especially the heterogeneity of system cores and modules imposes challenges for flexible on-chip clock generation. New solutions for these issues are developed in this work.

The scaling of modern complementary metal oxide semiconductor (CMOS) technologies [ITR11c] enables the integration of complex systems on a single chip. However, the capability of technology scaling with the number of producible devices on a single die can not be followed by the ability of designing and verifying such large systems. This phenomenon is called design productivity gap [AWZ08, Lev04, ITR11b]. Heterogeneous MPSoCs which split its total functionality over separated cores are one approach to this issue, since they allow re-use of design blocks (intellectual property (IP) cores) to ease and speed up system design implementation [Bra99]. System complexity is then achieved by scaling up the number of cores, which imposes new challenges for system architecture, software implementation and on-chip communication fabrics [Hen03, AWZ08]. Besides this, the main challenge of todays integrated circuit design is to limit the system power consumption which comes with increasing number of functionality and compute power on a single chip [ITR11c]. This can only be realized by application of advanced power management techniques [KFA+07].

Silicon integration of future complex MPSoCs requires an infrastructure environment which provides the basic functionality for clocking, on-chip and off-chip data communication as well as power management. In this work new concepts and circuits for clock generation within this MPSoC infrastructure are researched and developed. State-of the art clock generation circuits which are available as IP do not sufficiently cover requirements for advanced power management techniques and globally asynchronous locally synchronous (GALS) clocking architectures which require a large number of independent clock generators on a single chip. Often they are designed to serve a dedicated purpose, as for example processor core clocking or serial I/O link clocking [KMN+09]. They require large chip area because analog circuit components are used [HMY10] or consume lots of power for generation of ultra-low jitter clocks for a wide range of applications [TRF08].

To go beyond this, flexible clock generator solutions for heterogeneous MPSoCs are researched in this work as illustrated in Fig. 1.2. The generators shall be able to generate a wide range of output clock frequencies with low jitter and must be capable of ultra-fast frequency changes for advanced fine-grained power management. Therefore they must be instantiated *per-core* within the MPSoC which requires small chip area and low power consumption for reasonable costs and energy efficiency. In contrast to dedicated clock generator designs [KMN+09] concepts for clock generators that can be applied to a wide range of cores, thereby enabling efficient design re-use for implementation of complex multi-core systems, are to be developed here. With the introduction of new semiconductor technology nodes, efficient design implementation is critical to achieve a reasonable time-to-market. Therefore the clock generators are to be realized with as much digital content as possible to allow easy porting to new target technologies with state-of-the-art automated design implementation flows.

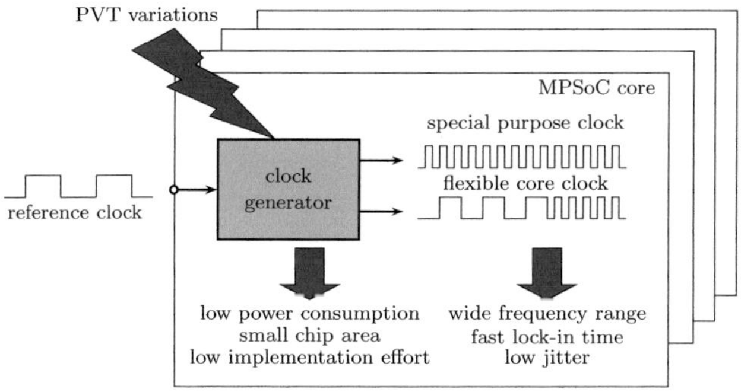

Figure 1.2: Flexible per-core clock generator for heterogeneous MPSoCs

The concepts and circuits that are researched and developed in this work are implemented into functional MPSoC demonstrator testchips where they are used to clock the system components. No dedicated stand-alone clock generator testchips are to be realized. This raises additional constraints for circuit robustness with

respect to process, voltage and temperature (PVT) variations, design verification, system integration and measurement access for testability but focuses this research to produce applicable results.
This research has been carried out as part of the "CoolBaseStations" and "Cool-RF-28" projects as part of the Leading-Edge Cluster "Cool Silicon"[1], in which the energy efficiency in information and communications technology (ICT) is to be increased [EMF+12, EMF+13].

This work is structured as follows. In Ch. 2 a general overview over clocking architectures, power management techniques and on-chip data transmission fabrics in modern heterogeneous MPSoCs is given, from which constraints for clock generation are derived. Also the testchips that contain the circuits of this work are introduced here. In Ch. 3 digitally controlled oscillators (DCOs) as key components of integrated clock generators are presented. A DCO in 65nm CMOS technology featuring a novel active current bias scheme and a DCO in an advanced 28nm CMOS technology node are presented. Differential buffers for DCO clock distribution are presented and a theory on their phase alignment properties is developed. In Ch. 4 ADPLLs concepts for clock frequency multiplication are presented and circuit realizations in 65nm and 28nm are shown. Based on a numerical system model, ADPLL architectures are analyzed and novel fast lock-in schemes are proposed. The aspect of ultra-fast frequency changes for MPSoC power management is solved by the open-loop clock generators which are presented in Ch. 5 and theoretically analyzed for their phase mismatch sensitivity. Ch. 6 summarizes the results of this work, shows application examples of the circuits that have been developed here and gives an outlook to further work.

[1] www.cool-silicon.de, The Leading-Edge Cluster "Cool Silicon" is sponsored by the German Federal Ministry of Education and Research (BMBF) within the scope of its Leading-Edge Cluster Competition

2 Multiprocessor Systems-on-Chip

As target application for the clock generator concepts and circuits of this work, this chapter gives a detailed overview over heterogeneous MPSoCs. State-of-the art concepts for clocking, power management and on-chip data transmission are introduced and advanced novel techniques for ultra-fast dynamic voltage and frequency scaling (DVFS) and energy efficient serial on-chip links, which have been developed within research activities closely related to the main content of this work, are reviewed. A core wrapper architecture is proposed here which encapsulates the functional MPSoC cores into a common infrastructure framework, including the clocking circuits. As result, key constraints for the clock generators as enablers for these new energy efficient MPSoC infrastructure components are defined as basis for the following chapters of this work.

2.1 Clocking Architectures

Clock signals as shown in Fig. 2.1 define the basic timing of sequential logic circuits which are edge sensitive or level sensitive with respect to the clock. The main property of a clock signal is its period $T = 1/f$. When two or more clocks have to be considered in an integrated system, the phase can be used to describe their relation. The phase difference between clock signals can also be considered in time domain, where it is referred to as skew t_{skew} [Fah05]. Non-idealities within the clock signal can be described by jitter, denoting the statistical timing variation of events (e.g. edges) or duty cycle distortion, describing the ration between logic high time $t_{\text{clk,H}}$ and the signal period T, which is ideally 50%. The clocking architecture of a MPSoC provides all sequential circuit parts with the required clock signals.

In *synchronous* designs all clocked elements receive a clock signal of equal period and defined phase. Synchronous designs can be implemented efficiently using state-of-the art synthesis and place&route flows, which can synthesize clock trees as shown in Fig. 2.2(a) automatically [Fah05]. But large synchronous clock trees can cause a significant part of the system power consumption [YB09]. The implementation effort increases when the synchronous circuit part increases and the realization of

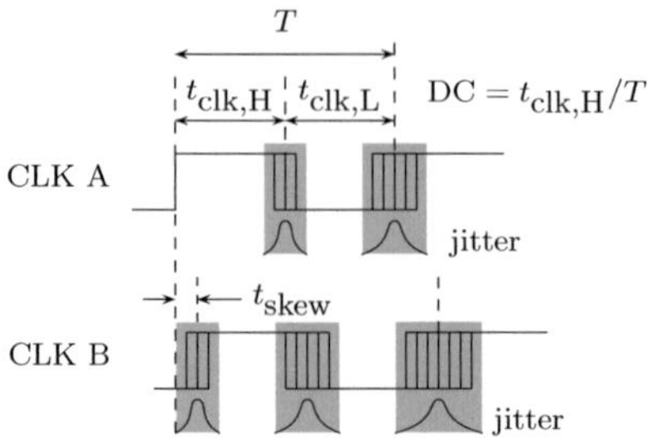

Figure 2.1: Clock signal definition

small clock skew is a challenge. Especially for high performance microprocessors, where large chip areas must be clocked synchronously at high frequencies clock skew directly translates into performance degradation (reduced speed) because jitter and skew are a higher fraction of the clock period. Clock skew reduction requires special circuit techniques, as for example synchronous global clock meshes as shown in Fig. 2.2(b) [SKDM10]. To reduce the power consumption of the clocking architecture resonant clock grids [HG12], where inductive elements are inserted in the clock mesh, are employed in state-of-the art X86 microprocessors [SAI+13]. However, fine-granular clock gating for power reduction is not possible here.
Since modern SoCs contain various components they include commonly several clock domains which are *locally synchronous*. The realization of dedicated clock domains for individual circuit components has a lot of benefits, as for example:

- Integration of circuit components running with different clock frequencies at the same time
- Different clock quality constraints for different circuit parts, e.g. low jitter clocks for I/O interfaces
- Application of power management techniques (see Sec. 2.2), e.g. power shut off, DVFS, adaptive voltage and frequency scaling (AVFS) [KFA+07]
- Easier timing closure for large multi-core integrated circuits, i.e. no large global synchronous clock trees required, independent optimization of smaller local clock trees possible.
- Reuse of pre-defined circuit hardmacros with internal clock trees for increased design implementation efficiency

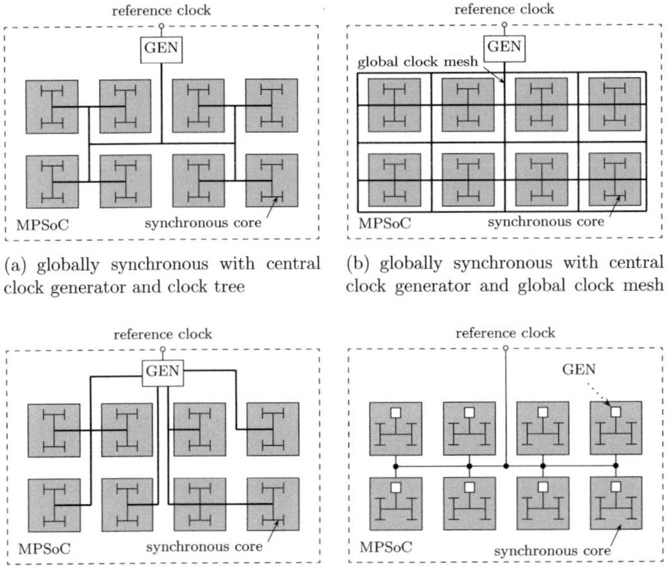

(a) globally synchronous with central clock generator and clock tree

(b) globally synchronous with central clock generator and global clock mesh

(c) globally asynchronous with central clock generator

(d) globally asynchronous locally synchronous with local clock generators

Figure 2.2: Clocking styles of MPSoCs, combinations of the shown architectures are possible

Based on the period and phase relation between the local clocks, different clocking styles can be distinguished. In *mesosynchronous* circuits each of the local clocks has the same nominal frequency but an unknown phase offset. In *plesiochronous* circuits the nominal frequency of the clock domains is the same, but a time varying phase shift can occur during system operation. Clock domains are considered to be *asynchronous*, if both clock period and phase are arbitrarily and can change during operation. This is referred to as GALS clocking architecture [YB09]. Based on the clocking style, techniques for data-synchronization between the clock domains must be employed [JAPR12]. Although GALS clocking architectures enable most benefits as listed above, as for example fine-grained per-core DVFS, they require the highest data synchronization effort. GALS systems have additional benefits in reducing the electromagnetic interference and the supply and substrate noise of MPSoCs [FKWG11].

The clock signals are typically generated by on-chip clock generators, which operate as frequency multiplier from an external reference clock signal (see Sec. 2.6).

Advanced microprocessors feature dedicated clocking solutions [KMN+09]. They include various clock generation circuits, which are directly adapted to their special requirements for each clock domain, e.g. for clock skew reduction, low jitter and good duty cycle. These circuits mainly are built up using special custom designed phase-locked loop (PLL) clock generators for core and interface clocking. The 80 core processor from [VHR+08] uses mesosynchronous clocking with a single PLL clock generator only, which does not allow individual per core frequency scaling. Also in heterogeneous MPSoCs commonly centralized clock generators are used which provide clocks for multiple cores or I/O interfaces [LWB+08] [RRH+11] as shown in Fig. 2.2(c).

In GALS systems local clock generators are used for each of the cores [KFG+11] as shown in Fig. 2.2(d). As example [Jip08] and [SLP08] use simple ring oscillators for the GALS cores, which are not suitable for DVFS or for generation of low jitter clocks at specified frequencies. Locally calibrated clock generators based on controlled delay lines are used in [MTC+00], but no fast core frequency switching is possible here which prevents fast DVFS performance level changes.

The application scenario which is covered by this work considers GALS MPSoCs which use local clock generator for each core and I/O module, that are capable to generate defined clock frequencies from a globally available reference clock signal, as shown in Fig. 2.2(d). This enables maximum flexibility with respect to clock generator design reuse, on chip communication (see. Sec. 2.3) and power management techniques (see. Sec. 2.2).

2.2 Advanced Power Management Techniques

The reduction of power consumption and thereby increasing the energy efficiency, which can be measured in Joule per computation, is essential for of complex MPSoCs. In high performance computing applications the power consumption is the main limiting constraint for system integration, because it directly effects the required power supply and cooling infrastructure. Increasing the energy efficiency of high performance computing systems with respect to the reduction of their CO_2 emission is currently addressed by many research projects, as for example within the Cool-Silicon cluster of excellence [EMF+12, EMF+13]. Besides that, mobile applications directly impose constraints for high energy efficiency of their electronic components to increase battery lifetime [AF11]. In addition to well established power reduction techniques, like clock gating, where the clock of non-active sequential circuit parts is disabled, or power shut off, where non-active circuit parts are disconnected from power supply [KFA+07], advanced techniques offer additional energy saving benefits for system operation. Two of those techniques are briefly presented in the following subsections and their requirements for flexible clock generation are highlighted.

2.2.1 Dynamic Voltage and Frequency Scaling

DVFS is a widely used power management technique to reduce the energy consumption of MPSoCs [MB10]. In general, the power consumption of an integrated digital circuit core can be expressed by

$$P = \underbrace{\frac{V_{\mathrm{DD}}^2}{T_{\mathrm{core}}} \cdot C \cdot a}_{\text{dynamic}} + \underbrace{V_{\mathrm{DD}} \cdot I_{\mathrm{leak}}}_{\text{static}} \tag{2.1}$$

where C denotes an equivalent capacitance of the core logic devices and interconnects and $a \in (0;1)$ is the effective logic toggle rate. From this, the energy consumption for a given core task is

$$E_{\mathrm{task}} = \left(\frac{V_{\mathrm{DD}}^2}{T_{\mathrm{core}}} \cdot C \cdot a + V_{\mathrm{DD}} \cdot I_{\mathrm{leak}}\right) \cdot t_{task} \tag{2.2}$$

$$= V_{\mathrm{DD}}^2 \cdot n_{\mathrm{task}} \cdot C \cdot a + V_{\mathrm{DD}} \cdot I_{\mathrm{leak}} \cdot T_{\mathrm{core}} \cdot n_{\mathrm{task}} \tag{2.3}$$

where n_{task} denotes the number of clock cycles per task. Commonly, the static leakage as second term in Eq. 2.1 and Eq. 2.3 is addressed by fine-grained runtime power gating during idle periods of the core during system operation. A significant reduction of the dynamic energy consumption is possible when scaling the supply voltage V_{DD}. As example, for a processor core in 65nm CMOS technology, energy per task reduction of up to 40% can be achieved [Zhe11]. At the same time the core clock frequency $1/T_{core}$ must be reduced to achieve timing error free operation under the low voltage conditions.

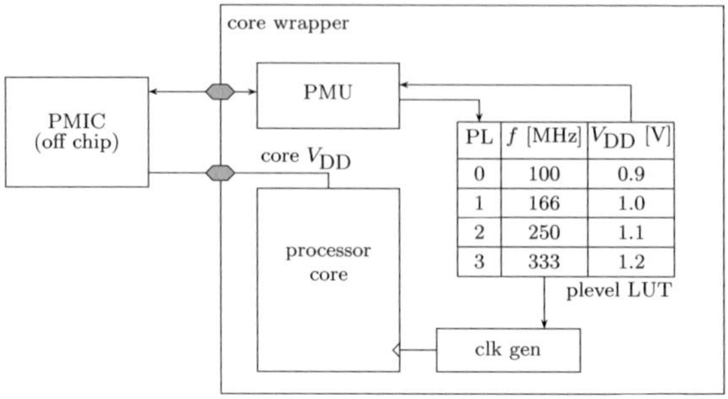

Figure 2.3: DVFS architecture block level schematic

A pair of a core frequency setting and a corresponding supply voltage is defined as performance level (PL). For DVFS the PLs are stored in lookup tables, as visualized in Fig. 2.3. These pairs are defined based on worst-case timing sign-off conditions to ensure safe system operation at each PL with respect to PVT variations. A power management unit (PMU) selects the PL for the active task based on its control input from the MPSoC core manager which is scheduling the task execution [AF10]. The frequency setting is fed to the core clock generator and the supply voltage value is fed to a voltage regulator, which in conventional implementations is often realized by off-chip power management ICs (PMICs).

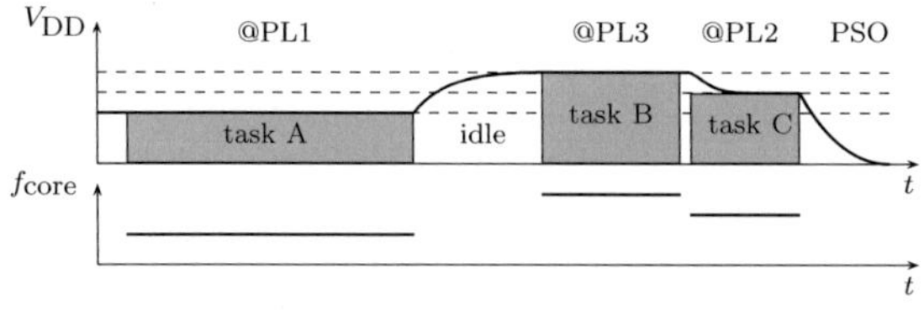

Figure 2.4: Task performance level change scheme of conventional DVFS

An example DVFS switching scheme is shown in Fig. 2.4. A first task A runs at a low PL. Then the PL is increased and the supply voltage is changed by reprogramming the external PMIC. Depending on the used interface (e.g. I^2C) this can take several tens of microseconds. During this time the exact supply voltage of the core is not

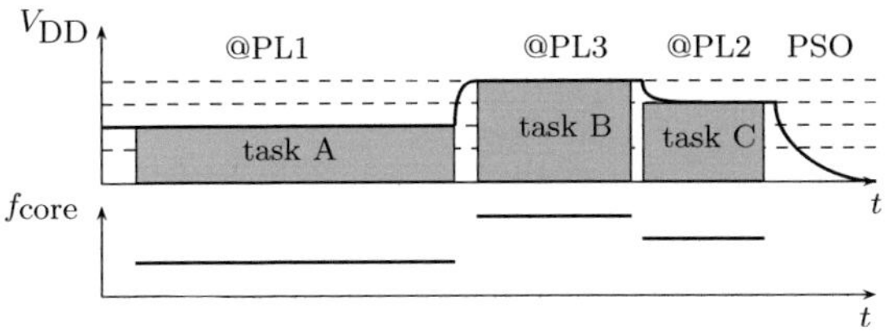

Figure 2.5: Task performance level change scheme of fast DVFS with multiple on-chip supply rails

known. Therefore the following task B must be delayed until the target supply voltage level has settled. Then the higher frequency clock signal can be applied to the core. This introduces idle times that limit the throughput of the MPSoC. For a PL reduction for task C, the lower frequency can directly be applied although the supply voltage has not yet settled.

The DVFS scheme can be applied within MPSoCs at different levels of granularity. In clustered DVFS schemes, as proposed in [KZS11], sets of MPSoC cores are combined for voltage and frequency scaling, but definition of this clustering heavily depends on the logic system design, because typical workloads have to be estimated during design phase. In contrast, fine-grained *per-core* DVFS enables individual PL scheduling for each MPSoC core during operation, which can help to optimize the energy efficiency of the integrated system. Using the conventional approach with external PMICs multiple programmable supply voltage domains would be required [SJJ+11], which leads to a significant increase of the chip pin count and the control effort to external PMICs. In [KGWB08] a solution for per-core DVFS with on-chip regulators using inductors connected to flip chip bumps is presented. However, this provides a significant influence on packaging. Another approach to per-core DVFS is to realize multiple supply voltage domains from off-chip regulators and to switch the cores individually between them. In [TCM+09] a DVFS architecture with two on-chip supply levels is presented. As visualized in Fig. 2.5 the switching of the core voltage between the on-chip rails is significantly faster and thereby reduces the core idle times during PL changes.

In [HSE+12] a power management architecture for fast fine-grained per-core DVFS in heterogeneous MPSoCs has been presented which has been developed in close context to this work. Its structural schematic is shown in Fig. 2.6. It contains a highly configurable PMU that controls the power switches connecting the processor core to the on-chip supply rails. Multiple core wrappers can be connected to the on chip power rails, as shown in Fig. 2.7. The PMU includes a configurable switch

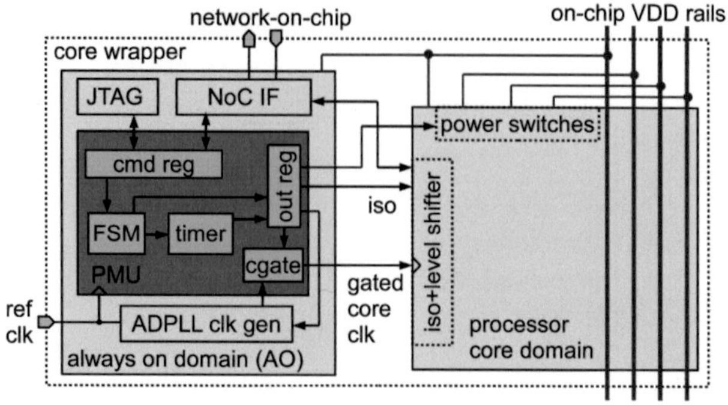

Figure 2.6: PMU core wrapper, [HSE+12]

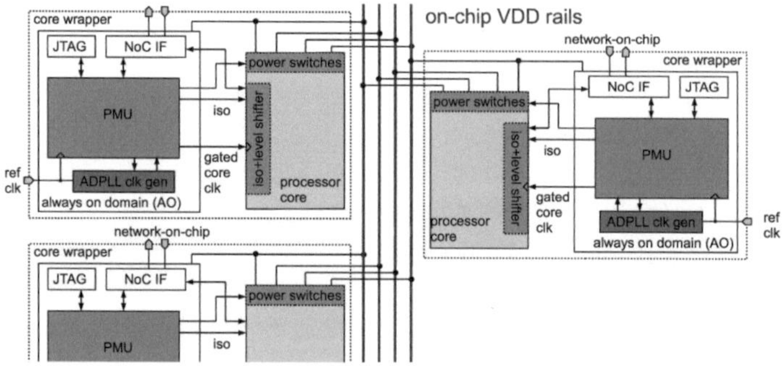

Figure 2.7: Core wrapper of multiple cores

scheduling scheme to reduce supply voltage switching noise [DS05] during power-up and PL changes of the cores by means of pre-charging techniques [SH06]. This is essential for safe operation of active cores while other cores on the same supply rail change their PL.

This power management scheme has been implemented in the "Atlas" testchip (see Sec. 2.8) in 65nm CMOS technology. It is used for power management of a vector DSP (VDSP) core. Ultra-fast PL level changes in time ranges well below 100ns can be achieved without disturbing active cores on the same supply nets. As example Fig. 2.8 shows the measured supply voltage of the core being switched and a second core on the target supply net. Using the implemented switch scheme a supply voltage change within 20ns is achieved.

This defines tough constraints for the core clock generator. First, a wide range of

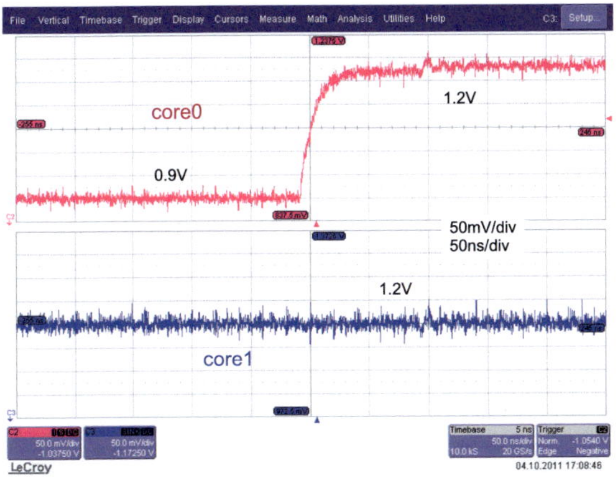

Figure 2.8: Measured voltage at PL change, "Atlas" testchip, 65nm, [HSE+12]

core frequencies must be supported to allow fine adaption of the PL frequency and supply voltage pairs for a variety of different cores with the heterogeneous MPSoC. Second, it must be capable to change the output clock frequency *instantaneously* to the specified value from the PL lookup table to enable ultra-fast DVFS as presented in [HSE+12]. The development of core clock generation concepts for instantaneous frequency changes and their circuit realization are in focus of this work.

2.2.2 Adaptive Voltage and Frequency Scaling

Fig. 2.9 shows the block level schematic of AVFS. In contrast to DVFS the supply voltage which corresponds to a certain core clock frequency is not predefined based on worst-case PVT assumptions but is determined during system operation by closed loop regulation [KFA+07]. Therefore hardware performance monitors (HPMs) are used to monitor the speed performance of the core logic. Commonly delay lines or ring oscillators as critical path replicas are used for this [ES04, Zhe11, INS+12]. Thus the AVFS scheme benefits from the fact that the supply voltage can adjusted to its (near-) minimum value for a given clock frequency and under consideration of the current PVT condition of the core, as monitored by the HPM. By this significant energy savings for task execution can be achieved (up to 27% [ES04], up to 40% [Zhe11], up to 45% [MPPdG04]). The AVFS technique can also be used to run cores with low supply voltage in the near-threshold or even sub-threshold range for ultra-low energy consumption [LJB+13].

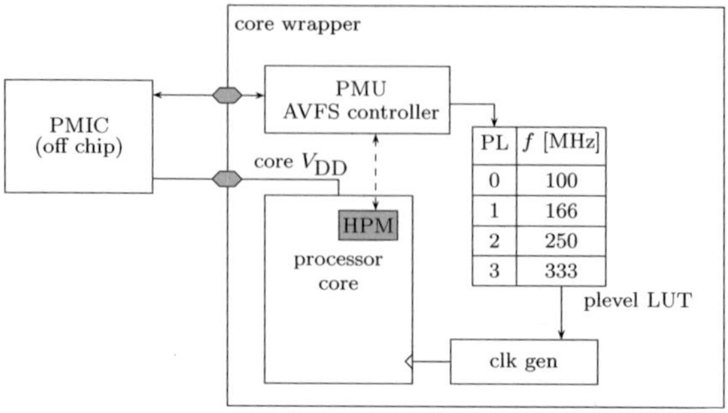

Figure 2.9: AVFS architecture block level schematic

Generally, the AVFS scheme can also benefit from clock generators with a wide range of output frequencies for fine granular performance level scaling. Also fast switching helps to reduce core idle times, e.g. when going from a higher frequency to a smaller one. In this case the clock frequency can be changed instantaneously and the core can continue its low frequency operation. The supply voltage then is regulated by the closed AVFS loop to its new target value.

2.3 Networks-on-Chip

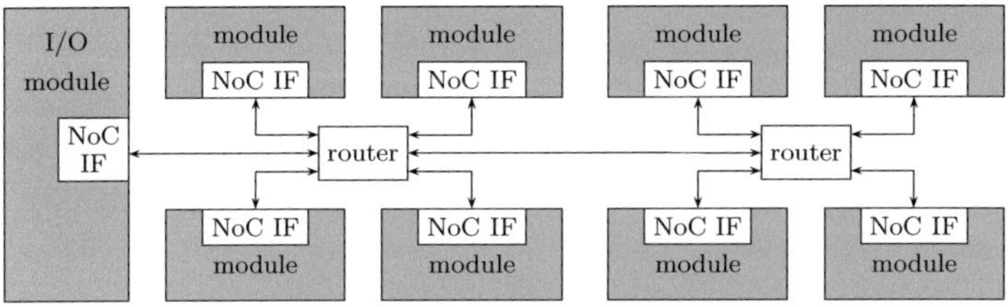

Figure 2.10: Example NoC structure

With increasing numbers of modules (e.g. cores, I/O components, on-chip memory) in modern MPSoCs, their data interconnection fabric is a main challenge for system implementation. It contributes to a significant part of the total on-chip energy consumption (i.e. in terms of energy per bit) and die area. Additionally it significantly impacts system performance (i.e. in terms of clock frequency) and throughput (i.e. in terms of data per time). But also quality-of-service (e.g. in terms of guaranteed latency) plays an increasing role within modern MPSoCs [Win11].

The network-on-chip (NoC) is a widely used communication architecture to address these challenges [AJ04, AWZ08]. The NoC connects the MPSoC modules in a packet based network, where data is transferred over routers from source to target destination. The routers and the modules are connected by point-to-point links. An example NoC topology is visualized in Fig. 2.10. The NoC approach provides significant flexibility especially in heterogeneous GALS MPSoCs with advanced power management, where the communication fabric can be completely abstracted from the cores and modules. A generic NoC interface can be integrated within each core wrapper, being independent from its logic content.

The optimization of NoC topologies with respect to energy efficiency [BMM07], throughput, quality-of-service and fault tolerance is in currently in focus of research activities [AJ04, Win11, Hof12].

2.4 Global On-chip Data Links

2.4.1 Overview

The performance in terms of throughput, latency and energy per bit of point-to-point connections within NoCs is limited by the physical circuit implementation of

these links. For short distance connections, conventional full-swing CMOS signaling of parallel data buses can be used efficiently, providing low implementation effort with acceptable performance.
However, the physical realization of some NoC topologies, as for example hierarchical NoCs [WPG10], require long global point-to-point in the range of some mm. Also the efficient floor plan design of an MPSoC can produce the need for long distance connections, as for example when I/O modules close to the chip edges (e.g. DDR2 interfaces, SerDes Links) have to be connected to the center area.
As the physical length of point-to-point connections increases, conventional CMOS signaling runs at its limits. The power consumption of the full-swing signals increases dramatically [MSK+10] and active buffers have to be inserted to reduce the delay and retain the signal slopes on the global lines with significant RC damping [NS10]. This additionally influences floor planning, because the insertion of active buffer cells must be possible for these long distance links. It prevents routing over pre-defined macro blocks, such as memory arrays or cores realized as hard macros. Also timing sign-off of global links is a challenge, especially in GALS architectures, where no global synchronous clock signal is available.
A suitable approach is the use of high-speed global on-chip links, where data is transmitted over long distances with high data-rates in the multi GBit/s range at low voltage swings. This enables low energy per-bit performance and no active buffers need to be inserted which significantly increases the flexibility for floorplan optimization. Data is serialized to higher rates per line to reduce the number of physical lines within the link [GIK+09]. The significant signal damping of RC-limited on chip lines can be circumvented by application of special capacitive driver techniques, which provide inherent pre-emphysis [HOH+08], [SHL+10]. Several circuit realizations of global low swing links have been reported [PKPF09] [SMK+09] [MSK+10], which mainly concentrate on the physical link without deeply considering clocking architecture demands for MPSoC integration.
The realization in [WHE+12], which employs clocking circuit components from this work, is the first complete high-speed NoC link transceiver, containing all circuit components required for transmission of a NoC packet over distances up to 6mm with up to 90GBit/s in 65nm CMOS technology. Its architecture is briefly presented in the following subsection.

2.4.2 High-speed NoC Link Architecture

Fig. 2.11(a) shows the block level schematic of the high-speed serial NoC transceiver architecture from [Wal10, WHE+12]. The transceiver provides an unidirectional data link for a 144-bit wide NoC packet from core (or router) A to core (or router) B, which are asynchronously clocked within the GALS scheme. The transmitter clock is provided by a local ADPLL clock generator, which drives the serializer logic. Data is serialized in portions of 16-bit to a double data rate (DDR) stream. Therefore 9 data slices are used to transmit the whole NoC packet. An asynchronous FIFO synchronizes the transmitted data from the core clock domain to the link clock domain.

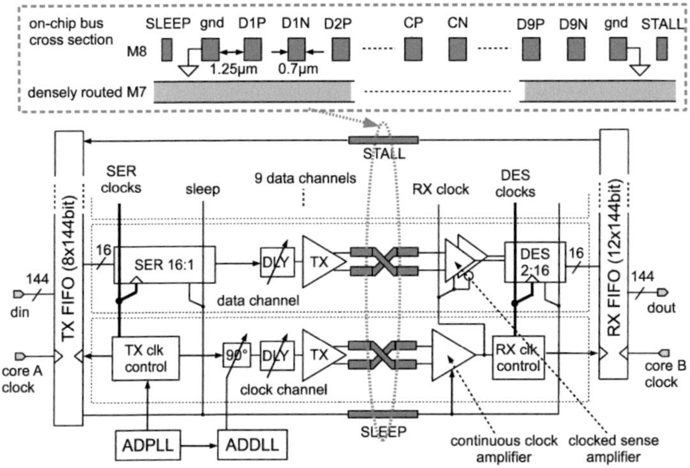

(a) block level schematic

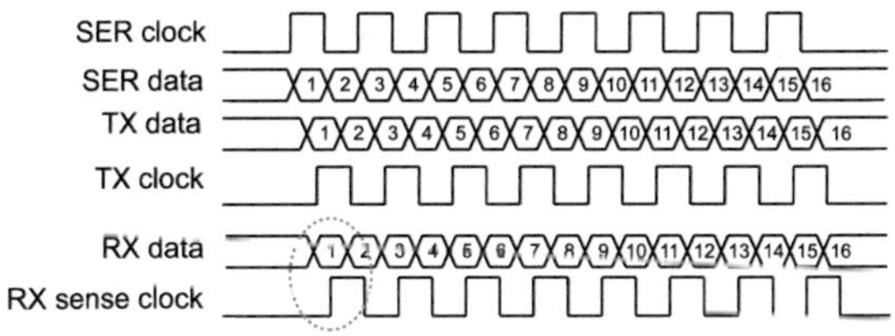

(b) signal waveforms

Figure 2.11: NoC Link Architecture from [WHE+12]

The transceiver uses a source-synchronous clocking scheme, where the link clock is transmitted over a similar low swing signaling channel as the data. This clock slice is shared among the 9 data slices which reduces the energy per-bit overhead of the forwarded clock. The clock is shifted by 90° at the transmitter to allow sampling at the middle of the DDR data eye at the receiver. The 90° delay cell is controlled by a delay-locked loop (DLL), which tracks PVT variations. One DLL can be shared among multiple NoC transceivers. The physical lines are driven by a combined capacitive and resistive line driver, realizing a low swing signal with amplitudes in the range from 100mV to 150mV and good signal eye opening at the end of the line. At the receiver side the clock signal is converted to full-swing CMOS level by a time-continuous amplifier. Clocked sense amplifiers are used to recover the data bits with high energy efficiency. Data is deserialized with the transmitted clock and synchronized to the receiver core clock domain by an asynchronous first in, first out (FIFO). A stall signal, driven by a static CMOS buffer from the receiver back to the transmitter, indicates if the receiver FIFO is almost full. This is used to stop the transmission until the packets are fetched from the receiver FIFO to prevent packet loss in the NoC.

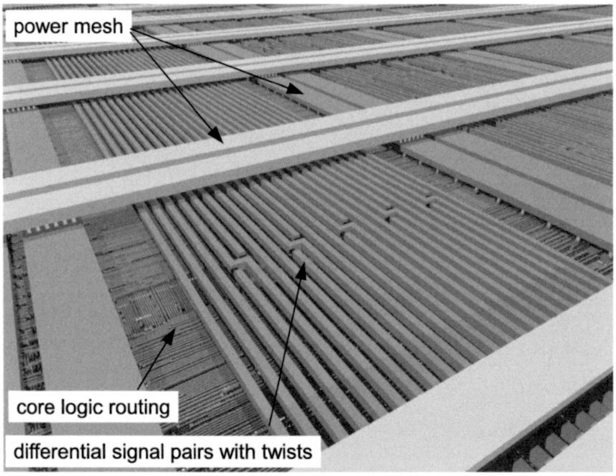

Figure 2.12: 3D visualization of global NoC link routing in the upper metal layers of the MPSoC

The physical data and clock lines are routed in the upper metal layers as visualized in Fig. 2.12. The lines are lying next to each other without additional shielding. Crosstalk to neighbor lines is minimized by insertion of twists in the differential lines using the scheme from [MSK+05]. The usage of the top metal layers within the chip power mesh allows to bridge pre-defined circuit macro blocks in the MPSoC. Additionally, the top metal lines with their higher thickness show less RC damping compared to the lower metal layers.

As a main benefit this architecture features completely stoppable clocking, where a clock edge is transmitted only if there exists a corresponding data bit, as shown in Fig. 2.11(b). This is mandatory for low power operation when no data is to be transmitted over the link. A dedicated sleep mode ensures that the time continuous clock amplifier is switched off during these idle periods. Thereby this circuit consumes no static idle power, except leakage. The sleep mode is enabled by the transmitter when its FIFO is empty. The sleep signal is driven by a static CMOS buffer to the receiver.

Since the transmitted data is sampled at the receiver by the same clock edge which has been used for transmission, this link architecture exhibits high jitter tolerance with respect to the ADPLL high speed clock. The time distance of the sampling clock edge at the receiver to its previous and following data transition is determined by the half clock cycle period (minimum distance between two edges in the DDR stream) and the delay of the 90° cell. Therefore it is only sensitive to *half-period jitter* of the transmitter clock and not to accumulated long term jitter of the ADPLL. If the absolute time of a clock edge shifts due to jitter accumulation (absolute jitter), the corresponding data bit is shifted accordingly. For details on jitter definitions see App. A.1.

Due to the fact that one clock lane is used to sample the received data of multiple data lanes, this architecture is sensitive to delay mismatch within the data lanes. Programmable delay elements are inserted in the transceiver to compensate this. An analysis of link yield reduction caused by delay mismatch and a compensation algorithm has been presented in [HWES10]. Within the manufactured circuit, the delay imbalances can be measured using an asynchronous sub-sampling technique as presented in [HWES11, HWES12].

As example, Fig. 2.13 shows some measurement results of this NoC transceiver implementation over 6mm distance in 65nm CMOS technology[1]. Fig. 2.13(a) shows the measured bit-error-rate (BER) for different combinations of supply voltage level and swing on the differential signal line at two different data rates. Fig. 2.13(b) shows the energy efficiency of the proposed link architecture for different data rates, when scaling the supply voltage such that there is a remaining 5% margin for a BER$< 10^{-12}$ as shown in Fig. 2.13(a). These energy measurements have been performed for different data toggle rates a, where the case $a = 0$ represents the clocking energy overhead of the transceiver. This shows that this NoC link architecture can provide low energy-per-bit on-chip signalling over long distances. Its speed versus energy performance trade-off is scalable by means of supply voltage and clock speed (data-rate) adjustment, similar to the DVFS scheme presented in Sec. 2.2.1.

The logic FIFO interface of this serial NoC transceiver is completely encapsulating

[1]"Atlas" testchip, see Sec. 2.8

the serialization circuitry, i.e. the physical link is invisible by the logic NoC. This allows to easily replace conventional parallel routings of the 144-bit NoC link, which are efficient for shorter distances, by the serial link for longer distances, without changes on the architectural NoC level.
Details on circuit implementation, delay calibration and measurement concepts can be found in [Wal10, HWES10, HWES11, HWES12, WHE^{+}12].

2.4.3 Clock Generators for High-speed On-chip Links

The serial NoC transceiver is driven by the clock of the transmitting core or router in the MPSoC. This imposes requirements for the local MPSoC clock generation circuits that are developed in this work. The local clock generator must be capable to provide a high-speed clock in the GHz range. In this work a maximum nominal frequency of 4GHz with a nominal data-rate of 8GBit/s per lane is specified[2]. It should be switchable between different frequencies to exploit the DVFS capability of the link. Good duty cycle of 50% is mandatory for DDR signalling on the NoC link. Due to the source-synchronous architecture, the jitter requirements of the serial on-chip links are relaxed, as explained in Sec. 2.4.2. Compared to I/O and memory interfaces, as for example DDR2, no accumulated jitter specification is required for the serial on-chip link clock.

[2]Overclocking of the serial link is possible by reprogramming the ADPLL frequency divider another value than the default one.

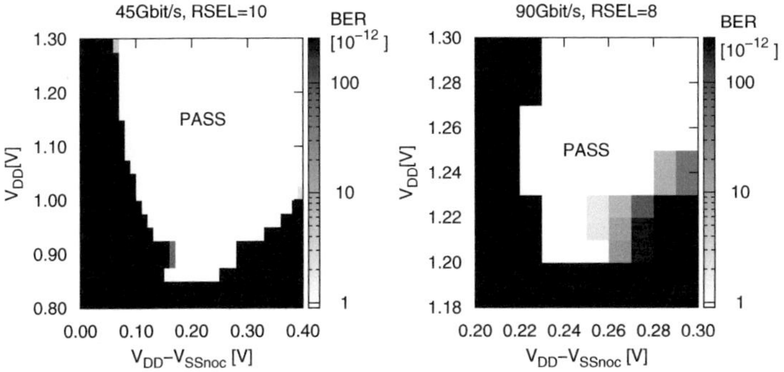

(a) BER shmoo plot, BER is measured for different signal swings on the link which are defined by the auxiliary supply voltage V_{SSnoc}, and the line driver parameter RSEL, [WHE+12]

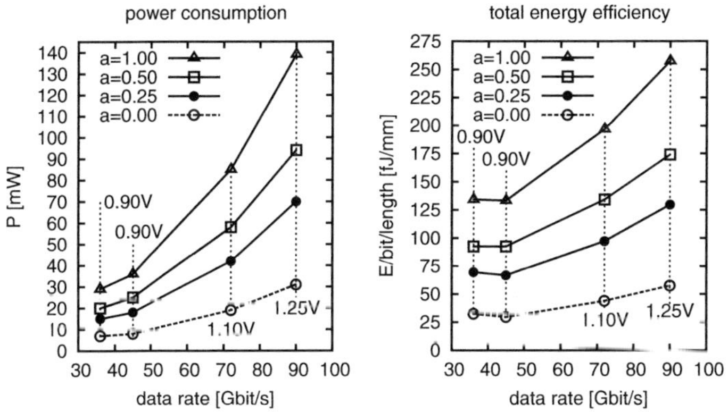

(b) power consumption and energy efficiency

Figure 2.13: NoC link measurement results, [WHE+12]

2.5 Core Wrapper

The circuit components for clock generation, interface to the NoC and the power management functionality of GALS MPSoC cores as presented in the previous subsections can be labeled as *infrastructure* components, because they are not directly related to the logic content of the core itself, which can be a RISC core, a DSP or a hardware accelerator for a specific algorithm for example. Also NoC routers with individual clock generators can be considered as cores in this approach.
Therefore a core wrapper, which encapsulates the logic core together with the common infrastructure components is a useful approach to enhance design efficiency for heterogeneous MPSoCs with various types of cores. This helps to increase design implementation efficiency by reuse of the common infrastructure IP components. The core wrapper is realized as highly parameterizable register transfer level (RTL) description which therefore allows technology independent reuse for a large variety of applications.

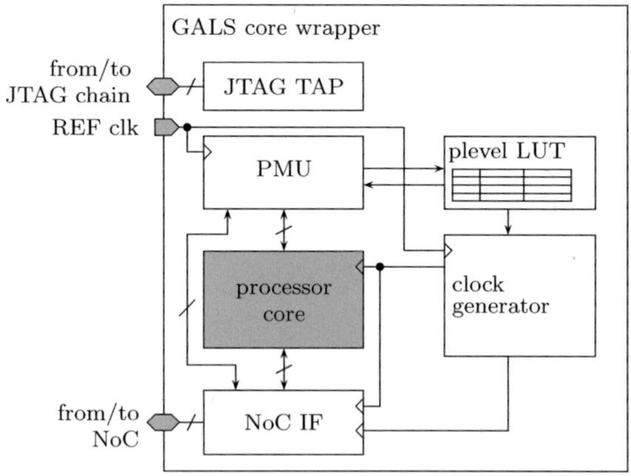

Figure 2.14: GALS MPSoC core wrapper

Fig. 2.14 shows the block level schematic of the generic core wrapper which is used in the MPSoCs addressed in this work. It consists of the following components:

- A NoC interface connects the core to a packet based NoC. The interface can be realized by conventional parallel connection of the NoC packet bits or using a high-speed serial on-chip transceiver as presented in Sec. 2.4.
- A PMU provides power management functionality in terms of power-shut off,

DVFS or AVFS to the core. For AVFS functionality an interface to HPMs can be included. The PMU is controlled by commands from the NoC.

- A versatile clock generator generates clocks for the core and (if required) for the NoC transceiver. This component in focus of the main part of this work.
- A Joint Test Action Group (JTAG) interface provides access to the core and the infrastructure components for test, configuration and debug purposes.

2.6 Local Clock Generators for GALS MPSoCs

As shown in Sec. 2.1, the GALS clocking architecture has many benefits, compared to globally synchronous clocking. However, it requires local clock generators that have to fulfill some general requirements:

- A wide range of clock frequencies must be provided for fine-grained power management techniques (e.g. DVFS, AVFS). The switching times between these output frequencies must be as small as possible to reduce idle times when changing the performance level of MPSoC cores.
- Special purpose clocks are required for interface clocking (e.g. DDR2/3) or high-speed network-on-chip data links. Special clock jitter specifications must be fulfilled [JED09], [JED10].
- The clock generator must be disabled for power gated cycles, with minimum static current consumption. The re-lock time after this off-state must be as small as possible.
- Low power consumption is mandatory to reduce the energy overhead of local clocking and to benefit from advanced power management (e.g. DVFS, AVFS).
- Small chip area is mandatory for per-core instantiation of the clock generator.
- The clock generator should be easily portable to another semiconductor technology node, to reduce design implementation time. The number of custom-designed circuit blocks should be minimized and as much as possible content should be realized as digital circuit to benefit from both the fast digital RTL-to-GDS implementation flow and the excellent scaling of digital logic cells in smaller technology nodes.

Ideally, one clock generator circuit should be capable to fulfill all requirements mentioned above. Previously published clock generators are only capable to fulfill parts of these requirements. In [Jip08] and [SLP08] simple ring oscillator clock generators are used. They are not suitable for DVFS frequency switching and do not generate low jitter clocks at defined frequencies. The locally calibrated clock generators based on controlled delay lines in [MTC$^+$00] do not allow fast switching between clock frequencies. The Flying Adder frequency synthesizer presented in [Xiu07] can generate a wide range of frequencies with low jitter but requires large chip area.
Therefore this work attempts to realize a clock generator for heterogeneous GALS MPSoCs, which can fulfill the explained requirements and specifications. It should be instantiated *per-core*. Thereby low power consumption and small chip area are the key performances to be optimized. The required output frequencies are in the range from below 100MHz (for processor cores at low performance levels) to up to some GHz (for high-speed network-on-chip signaling). Those have to be generated from a global reference clock which is typically in the range of some 10MHz.
For this clock frequency multiplication task, PLLs [Fah05] or DLLs [KKK$^+$06] are widely used. Their circuit structures are illustrated in Fig. 2.15.

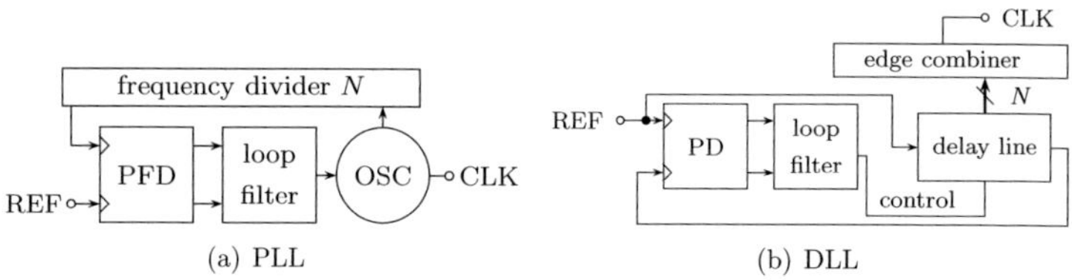

Figure 2.15: Circuit structures of PLL and DLL clock frequency multipliers

In a PLL as shown in Fig. 2.15(a), a controlled oscillator provides a period T_0. A loop frequency divider divides the oscillator signal frequency by N, realizing a divider output period of $N \cdot T_0$. This is compared with the reference clock period T_{ref} at the phase frequency detector (PFD). If the phase or frequency differ, the tuning signal of the oscillator is adjusted accordingly. The low-pass loop filter ensures stability of this control loop. By programming of the loop division ratio N, a wide range of output frequencies can be generated.
In a DLL as shown in Fig. 2.15(b), the reference clock signal with period T_{ref} is fed to a delay line of N_{tap} equal elements with tunable delay. The output of this delay line is connected to a phase detector together with the un-delayed reference clock. The delay line is tuned based on the phase detector output such that its total delay

equals T_{ref}. Then each delay element has a delay of $T_{\text{ref}}/N_{\text{tap}}$, thereby providing a multi-phase representation of the reference clock period with N_{tap} phases at their outputs. A high frequency clock is generated from combination of these multiple phases ([LCL09], [KKK+06]). The smallest output period to be generated from this delay line is $2 \cdot T_{\text{ref}}/N_{\text{tap}}$ [vdBKVN02].

A detailed comparison between PLLs and DLLs for low jitter frequency multiplication has been presented in [vdBKVN02]. PLLs tend to show larger long-term jitter (see. App. A.1) because the device noise in the oscillator accumulates in the oscillation loop [Fah05]. This is especially critical for fulfillment of the DDR2/3 memory interface clock specification. Jitter accumulation is not present in DLLs [KKK+06]. Their main jitter sources come from the reference clock and the noise of the controlled delay elements. It has been shown in [vdBKVN02] that a PLL based clock generator can generate a lower period jitter output clock with the same power budget compared to a DLL based circuit, because less delay stages are required in the PLL oscillator compared to the DLL delay line. Additionally DLL based frequency multipliers suffer from mismatch in the multi-phase combiners which results in increased period jitter [vdBKVN02]. If wide frequency ranges are required (as in this work), a DLL based solutions would have a large number of delay cells and a high logic effort for the frequency multiplication logic.

Therefore, in this work a PLL based clock multiplication solution is employed. Fig. 2.16 shows the block level schematic of the local clock generator. The PLL multiplies the reference clock frequency by a factor of N and provides a multi-phase output signal with the period $T_0 = T_{\text{ref}}/N$. The target period of T_0 is chosen such that a robust circuit implementations of the oscillator and the PLL loop divider can be realized in the target technologies (65nm and 28nm CMOS). This especially includes PVT variations. From this PLL output clock lower core frequencies are generated by open-loop frequency division and higher frequencies are generated by open-loop frequency multiplication (see Sec. 5.1). The smaller required multiplication factor keeps the clock multiplication logic simple compared to purely DLL based solutions with wide multiplication factor ranges [KKK+06]. The advantage of this technique is, that changes of the output frequencies can be realized by re-programming the open-loop clock generators without time-consuming re-locking of the closed loop PLL.

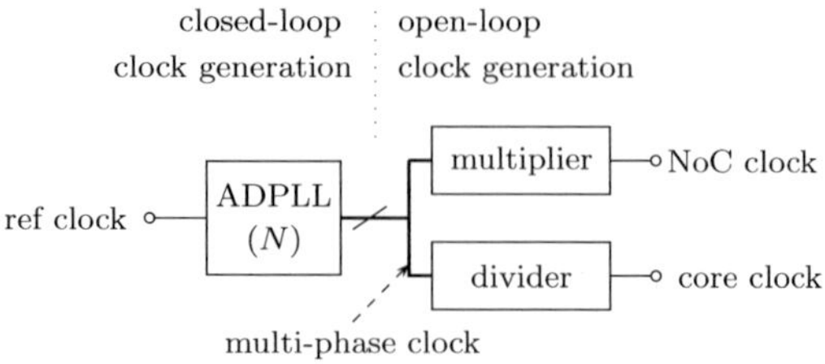

Figure 2.16: Local clock generator block level schematic

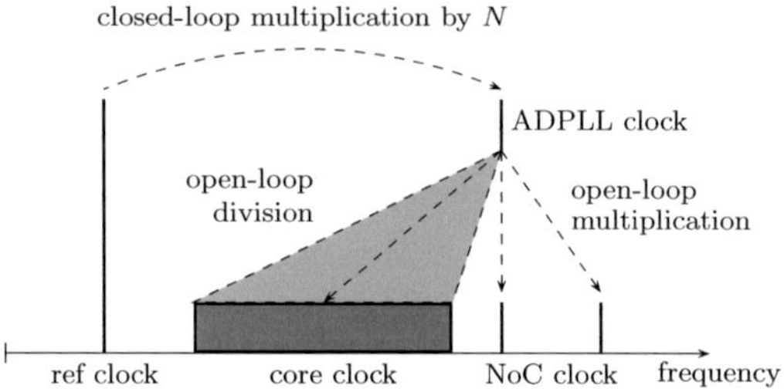

Figure 2.17: Frequency scheme illustration of the local clock generator

Fig. 2.17 illustrates the frequency plan of the local clock generators developed in this work. The core clock and NoC clock output periods read

$$T_{\text{core}} = \frac{T_{\text{ref}}}{N} \cdot N_{\text{olclkg}} \tag{2.4}$$

$$T_{\text{NoC}} = \frac{T_{\text{ref}}}{N} \cdot \frac{1}{N_{\text{NoC}}} \tag{2.5}$$

where $3 \leq N_{\text{olclkg}} \leq 24$ and $N_{\text{NoC}} \in 1, 2$ as presented in Sec. 5.1. The PLL loop division ratio N is static during system operation but can be adjusted for adaption to other reference clock frequencies or realization of different sets of core frequencies for different instantiations of the same local clock generator. Unless not stated otherwise, the PLL parameters as summarized in Tab. 2.1 are used in the following.

Table 2.1: Local clock generator parameter summary

	comment	**value**
T_{ref}	reference clock period	20ns
T_0	nominal PLL oscillator period	0.5ns
N	PLL loop division ratio	40

2.7 Silicon Implementation

2.7.1 CMOS Technology

Complex MPSoCs heavily benefit from the geometry shrink of modern CMOS technologies [ATE+09, WLL+09] with respect to both logic area reduction by lateral scaling and reduced power consumption by lower supply voltages [ITR11c]. Especially on-chip memory (e.g. static random access memory (SRAM)) which occupies a significant portion of the chip area of MPSoCs shrinks well with smaller technology nodes.

The circuits that are in focus of this work are targeted for implementation in state-of-the art CMOS technologies. The testchip realizations as summarized in Sec. 2.8 are implemented in a digital low-power 65nm CMOS technology, and a high-k metal gate 28nm CMOS process [ATE+09].

These CMOS technologies offer various types of devices. This includes transistors for core voltage domain as baseline devices and I/O transistors for different supply voltage options. Usually radio frequency (RF) and mixed-signal options of digital CMOS processes offer additional active and passive devices. Commonly core transistors are available in multiple threshold voltage (V_{th}) flavors, to allow the design trade-off between switching speed and static leakage power in multi-V_{th} implementation flows [KFA+07]. However, most of the additional devices causes the requirement for additional masks which increases production costs and restricts the re-use of the circuit component to those chips which use this corresponding process option. Therefore the clocking circuits which are developed in this work use the core transistor devices only. Additionally some poly resistors are used in the bias circuits of the 65nm DCO circuit presented in Sec. 3.2. Also the multi-V_{th} options of the target processes are used for design implementation, where low threshold voltage (LVT) or even super low threshold voltage (SLVT) devices are mainly used in the high-speed circuits of the ADPLLs. This reduces the active dynamic power consumption since smaller devices with less parasitic capacitance can be used when the low threshold option is chosen. The increased leakage is no issue here because these clocking circuits are small in terms of gate count and show high toggle rates. So dynamic power consumption dominates.

For design implementation, verification and characterization, the device models which are provided with the process design kits (PDKs) are used. The complexity of modern transistor models including various effects (short channel effects, temperature dependency, layout dependency (e.g. well proximity), noise) [Bha09], makes it impossible to use analytical models for exact design sizing. However, simple metal oxide semiconductor (MOS) transistor models [SN90, Uye01, Bak05] are used to analyze and explain basic circuit concepts. The statistical variation of process parameters is included in statistical models that can be used for Monte-Carlo circuit simulations. Global process variations that effect the delay of logic gates are implemented in corner model cards, typically representing the $\pm 3\sigma$ corners of the parameter sets. These are used for design verification and characterization by circuit simulation. Mismatch analyses are performed by Monte-Carlo simulations. In deep-submicron CMOS technologies on-chip interconnects are realized in the copper metalization layers. Although modern processes offer ultra-low k dielectric stacks [ATE$^+$09], the parasitic influences of the metal interconnects have significant impact on circuit performance because the aggressive geometric technology shrink reduces both distance and line width of the interconnect wires. This increases both parasitic coupling capacitances and series resistance. Therefore parasitic RC extraction is used to estimate the post-layout parasitics and to include them into the circuit netlist for final simulation and circuit characterization.

2.7.2 Design Flow

The clock generators which are researched in this work can be considered as mixed-signal circuits. From the signal perspective especially the DCOs contain both time and value continuous signals, e.g. programmable currents in the 65nm DCO realization as presented in Sec. 3.2. From the implementation point of view these components are optimized on transistor level, and in contrast to custom digital designs, analog building blocks (e.g. current mirrors) and devices (e.g. resistors) are used (see. Sec. 3.2). However, a large portion of the clock generators is realized as digital circuits. Especially those components running at high frequencies are optimized on transistor level in a custom digital flow. It is desired to realize as many circuit components as possible in the digital part in order to benefit from technology scaling of digital circuits and to reduce the design implementation effort by automated synthesis and place&route flows wherever possible.
Fig. 2.18 shows the mixed-signal design implementation flow that is used for the clock generators in this work. The analog and custom digital components are realized by manual design implementation on schematic and layout level within the custom flow. This includes verification by circuit simulation and layout verification by design rule check (DRC) and layout versus schematic (LVS). Various design and abstract views are generated for each custom circuit block which is directly integrated into the digital synthesis and place&route flow. This includes layout abstract, behavioral models for digital simulation and timing .lib files.

The digital circuits (e.g. ADPLL controller) and the toplevel of the clock generators are realized as RTL description. Functional verification is performed by digital and mixed signal circuit simulation. The clock generators are implemented using a standard digital synthesis and place&route flow including sign-off verification (e.g. static timing analysis (STA)). From this, an interface logic model (ILM) is generated which can be seamlessly integrated into the MPSoC toplevel design and implementation flow. The electronic design automation (EDA) tools that are used in this work are summarized in in App. A.2.

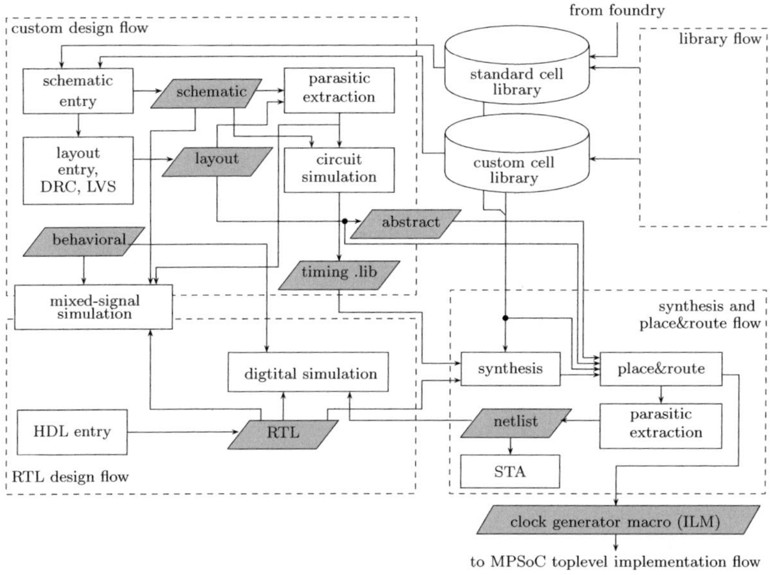

Figure 2.18: Mixed-signal design flow, simplified, PDK design resources not shown

2.7.3 Logic Cell Design

Digital circuits are efficiently implemented using standard cells which are provided as library including all required design views [Uye01]. Also in custom digital designs, the application of library cells eases schematic and layout implementation.

However, the addition of customized cells to the libraries which are tailored for a specific target application can significantly improve the performance of digital circuits, as for example in terms of higher speed of reduced power consumption [DC04]. For these cells all required views, including timing and power characterization, are generated to allow seamless integration into the digital implementation flow. As example, in [UHES10] a cell-based design register file for an single data rate (SDR) baseband processor is presented.

The main optimization target of customized digital cells for usage in high-speed clock generation circuits in this work is speed and special functionality. The logic cells that process clock signals are optimized for equal delays for rising and falling signal edges, by selection of the width ratio between the pull-up p-channel metal oxide semiconductor transistor (PMOS) and pull-down n-channel metal oxide semiconductor transistor (NMOS) devices in the gate. This is in contrast to data cells, which are optimized for the minimized sum of rising and falling edge delays.

All sequential elements are realized as static CMOS sequential logic [Uye01], which enables robust realizations even in small CMOS technology nodes. However, in the high-speed serialization and deserialization circuits for high-speed on-chip communication as shown in Sec. 2.4 contain dynamic latches [WHE$^+$12]. They enable high operation frequencies at low power consumption, but additional circuit effort is required to handle the performance degradation due to leakage in deep-submicron CMOS technologies. The trade-off between the speed of logic cells and their density can be adjusted by selection of the cell height (measured in routing tracks) and thereby the definition of the maximum drive strength per cell with. The high speed cells in this work are 14 tracks in height. Different standard cell libraries are used for implementation of the custom and semi-custom circuit parts as shown in Fig. 2.18.

For the 65nm implementations, a standard cell library provided from the foundry[3] is used for the core controller logic of the ADPLL clock generator as presented in Sec. 4.2. A 14 track high speed library with approximately 130 cells in 3 V_{th} versions has been developed at the Chair of Highly-Parallel VLSI-Systems and Neuromorphic Circuits for this 65nm technology. These cells are applied in the high speed circuit parts (e.g. DCO, frequency divider) of the ADPLL as well in the high-speed on chip links as shown in Sec. 2.4. An example cell layout is shown in Fig. 2.19(a). The 28nm testchips are realized completely by standard cells developed at the Chair of Highly-Parallel VLSI-Systems and Neuromorphic Circuits. This 14 track library contains approximately 200 cells in 4 V_{th} versions. An example cell layout is shown in Fig. 2.19(b).

For efficient custom design implementation it is desired to implement as much as possible cells within the fixed standard cell layout grid, which defines the location of power rails and allows easy placement of these cells. Especially in modern CMOS

[3]TSMC, 65nm LP CMOS, core standard cell library, 9 track height

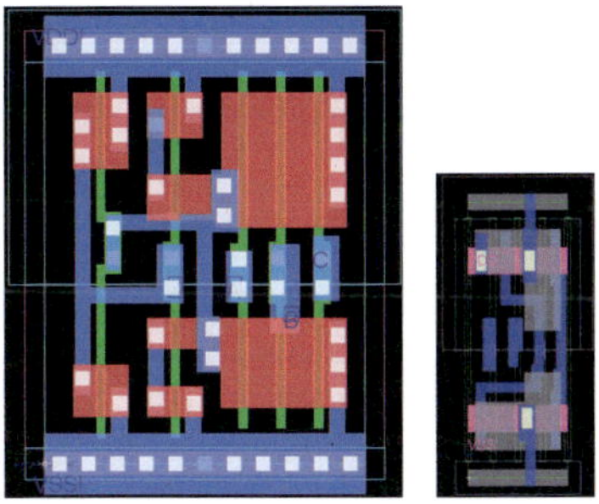

(a) 65nm (b) 28nm

Figure 2.19: Standard cell layout examples, not to scale

technologies, where complex layout design rules must be fulfilled for the frontend layers, this approach significantly eases layout generation, because after placing the pre-defined cells only routing in metal layers has to be performed. In the 65nm DCO and circuits as presented in Sec. 3.2 some analog components of a bias circuit and a digital-to-analog converter (DAC) are layed out in an analog fashion with free placement of MOS devices. In contrast, in the 28nm clock generation circuits as presented in Sec. 3.2 and Sec. 4.3 all circuit components are implemented on the standard cell grid.

2.7.4 Methods for Analog Custom Design

The custom design of analog and mixed-signal circuits imposes additional challenges because no completely automated toolflows for implementation are available, compared to the digital synthesis and place&route flows. There exist various numerical optimization methods for analog sizing that rely on time consuming circuit simulations [Gra07], but can realize robust circuit implementations that not only fulfill the nominal specifications but also show improved parametric yield. However, to allow efficient design reuse for analog SoC components (e.g. voltage references, voltage regulators, sensor interfaces) improved strategies for automated technology porting and automated sizing have to be developed [BNS+11], which speed up the implementation process. One challenge is the handling of complex semiconductor device models which have a wide range of parameters especially for small technology nodes, and can differ for different technology nodes and semiconductor foundries. These issues have been addressed within the SyEnA project [4]

In [HGH+10] a lookup table based flow is proposed where the key parameters of MOS devices with respect to analog circuit implementations (e.g. g_{m}) are stored in a generalized lookup table format, which makes them usable within novel automated analog sizing flows [BNS+11] that do not primarily rely on time consuming analog circuit simulations for optimization. The usage of lookup tables can make there sizing flows independent from the foundry specific PDK models or technology nodes. Additionally the fulfillment of constraints is essential to realize robust analog circuits [MGS08]. One important constraint for linear analog circuits is the saturation criterion of MOS transistors $V_{\mathrm{DS}} > V_{\mathrm{GS}} - V_{\mathrm{th}}$. But especially in smaller technology nodes with reduced maximum supply voltage the available signal voltage ranges which fulfill these constraints are limited. This results in the fact that proven circuit topologies (e.g. cascode current mirrors) can not be applied to analog circuit realizations in nanometer technologies. The prediction of the feasible voltage ranges of a given circuit topology for a target technology is mandatory to speed up the design implementation process and to avoid optimization effort on infeasible topologies.

In [HHSG10, GHH+11] a method for fast analysis of the feasible voltage ranges of analog CMOS circuits is presented. The MOS devices are replace by their linearized operating point (LOP) representation which is valid in the nominal DC operating point (DCOP) which is determined by a single DC simulation. The LOP equivalent schematic is shown in Fig. 2.20. The LOP parameters (linearization coefficients) are stored in lookup tables [HGH+10] and are provided as technology specific resource

[4]The SyEnA project (project label 01M3086) is supported within the Research Programme ICT 2020 by the German Federal Ministry of Education and Research (BMBF).

to the analysis tool. Due to the fact that the voltage relations between the MOS transistor nodes are purely linear, and the saturation constraints are linear inequalities as well, their valid regions with respect to the circuit input voltage nodes (e.g. analog input, supply voltage) can be determined by linear matrix analysis methods. They can be executed very fast and do not require circuit simulations.

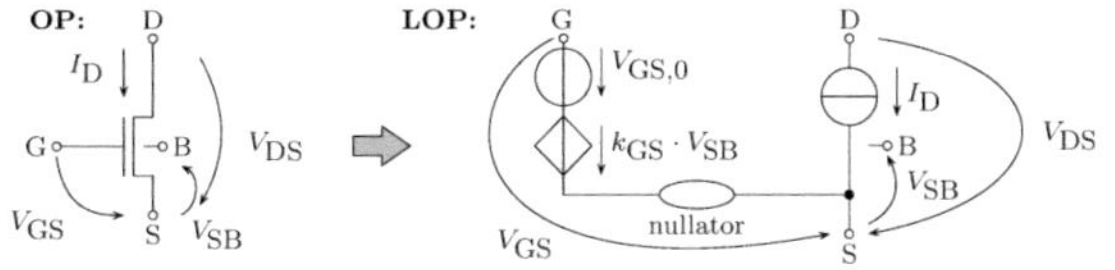

Figure 2.20: MOS transistor LOP model

As example Fig. 2.21 shows the LOP analysis results of a simple operational transconductance amplifier (OTA). It contains three high resistive input voltage nodes, which potentials have to be defined by the surrounding circuitry (supply voltage V_{DD}, input V_{in}, output V_{out}). The feasible voltage ranges of the supply voltage and the input signal are plotted, defining the available signal headroom for a given supply level. The LOP results (lines) are in excellent results compared to multiple DC sweep simulation runs (gray areas) but require only a single DC simulation.
Thereby very fast decision on the feasibility of a given topology for a target supply voltage level is possible and extremely useful in deep-submicron CMOS technology nodes. For example this method can be applied to predict the minimum supply voltage for the active DCO bias circuits, containing current sources and simple amplifiers, as presented in Sec. 3.2 of this work.

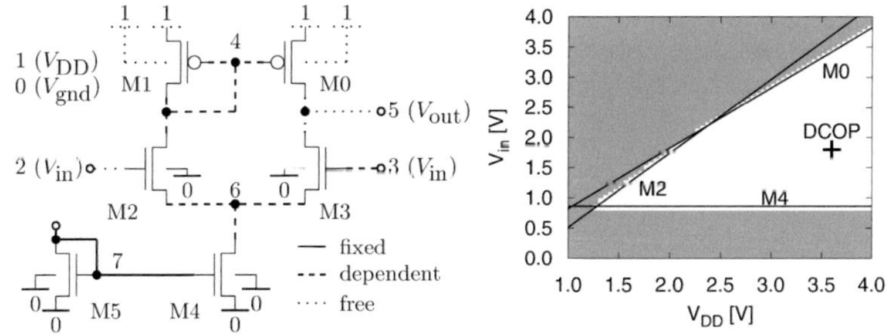

(a) OTA schematic (b) results ($V_{in} = V_{out}$)

Figure 2.21: Example LOP analysis, Spectre DC sweep with 50mV steps (gray fields: constraints violated) compared to LOP solution from single DCOP (lines)

2.8 Testchips

The clocking concepts and circuits that have been developed in this work have been verified by three testchips. They have been designed and implemented by the Chair of Highly-Parallel VLSI-Systems and Neuromorphic Circuits[5] at Technische Universität Dresden, focusing on the MPSoC infrastructure circuits and the physical implementation. The logic system architecture, NoC and processor cores of "Tommy" and "Atlas" have been designed by the VODAFONE Chair of Mobile Communication Systems[6] at Technische Universität Dresden. All silicon measurement results presented in this work have been obtained from these chips.

"Tommy", TSMC 65nm LP CMOS The "Tommy" demonstrator chip is the first silicon component prototype of the heterogeneous MPSoC architecture and infrastructure, developed within the CoolBaseStations project [EMF+12, EMF+13]. It includes two cores for hardware acceleration of mobile communication algorithms (FEC and sphere decoder) [WKA+12], which are connected by a packet based network-on chip. An FPGA interface based on high-speed source synchronous low voltage differential signaling (LVDS) links allows high throughput I/O. The GALS clocking architecture for the cores, the FPGA interface and the NoC routers are clocked by 8 ADPLL clock generators [HEH+13], as presented in Sec. 4.2 of this work. Furthermore "Tommy" contains three test links for high-speed serial on-chip communication (NoC) links with up to 72GBit/s over 6mm distance as presented in Sec. 2.4 with on-chip measurement and delay calibration capabilities [HWES10, HWES11]. Its die photo and block level schematic are shown in Fig. 2.22 and Fig. 2.23, respectively.

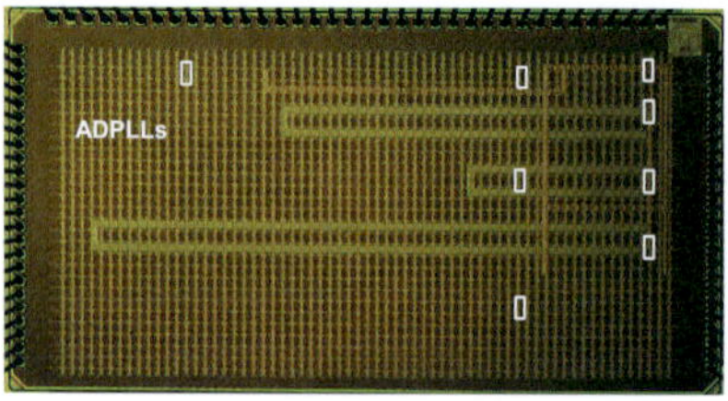

Figure 2.22: "Tommy" chip photo, 3.7mm x 1.8mm, 65nm CMOS, [WKA+12], positions of ADPLL clock generator marked

[5] http://hpsn.et.tu-dresden.de
[6] https://mns.ifn.et.tu-dresden.de

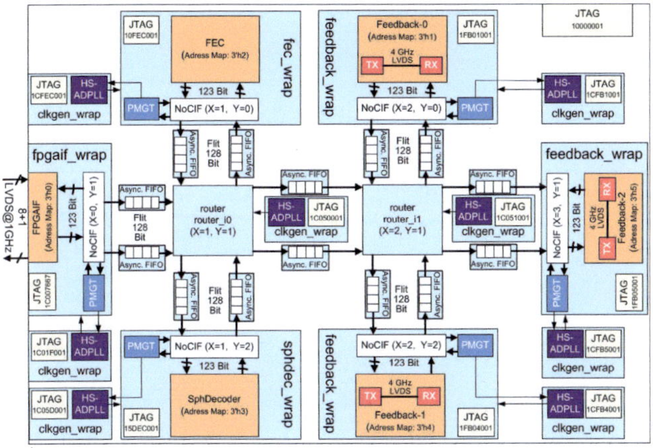

Figure 2.23: "Tommy" block diagram [Win10]

"Atlas", TSMC 65nm LP CMOS "Atlas", the second CoolBaseStations [EMF+12, EMF+13] testchip, contains two vector DSP cores, which are enabled for ultra-fast DVFS [HSE+12] as shown in Sec. 2.2. It includes an improved DCO within its 5 ADPLLs, featuring advanced compensation of temperature and supply voltage variations [HHH+12], as presented in Sec. 3.2. A high-speed NoC point-to-point testlink achieves 90GBit/s data-rate over 6mm uninterrupted on-chip interconnects [WHE+12]. An interface to external memory is realized by a DDR2 PHY (Synopsys® IP) and a source synchronous LVDS link is used for FPGA communication. Both I/O interfaces are clocked with ADPLLs from this work. The GALS cores and interfaces are connected by a packed based NoC, where one functional point-to-point connection over 1mm is realized by a serial NoC link with 36GBit/s data rate. The die photo and block level schematic are shown in Fig. 2.24 and Fig. 2.25, respectively.

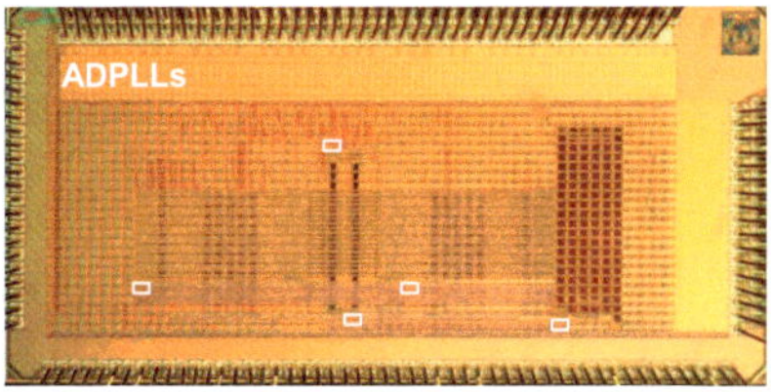

Figure 2.24: "Atlas" chip photo, 3.7mm x 1.8mm, 65nm CMOS, [WHE+12], positions of ADPLL clock generator marked

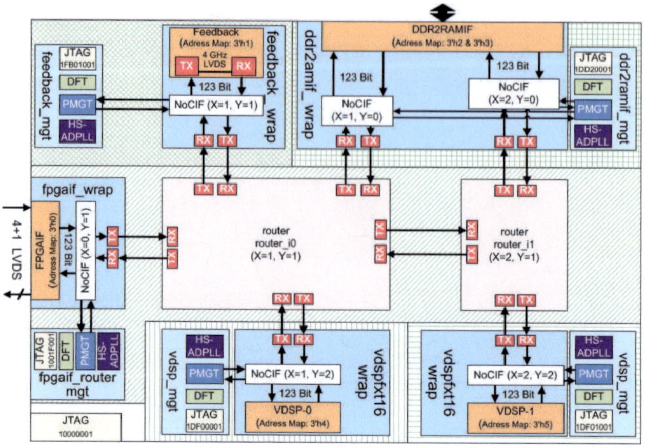

Figure 2.25: "Atlas" block diagram [Win10]

"Cool28SoC", GLOBALFOUNDRIES 28nm SLP CMOS The "Cool28SoC" is a testchip for low-power MPSoC circuit components in the state-of-the art 28nm SLP CMOS technology from GLOBALFOUNDRIES. It is built up completely using the in-house designed base IP of the Chair of Highly-Parallel VLSI-Systems and Neuromorphic Circuits, including standard cells, in total 96kByte SRAM macros as well as low-speed and high-speed I/O cells. A DSP from Tensilica® is used as processor core, which is enabled for AVFS, as presented in Sec. 2.2. The core clock is generated by an ADPLL clock generator developed in this work as presented in Sec. 4.3. Fig. 2.26 shows its layout and a partial die photo.

Figure 2.26: "Cool28SoC" layout and partial chip photo, die size 1.5mm x 1.5mm

2.9 Summary

A top-down overview from architecture to silicon implementation on the MPSoCs which are target systems for the clocking circuits being developed in this work has been given. From general architecture specifications and comparisons the GALS clocking architecture has been pointed out to be well suited for the application of advanced power management techniques. A novel ultra-fast DVFS architecture has been shown. The NoC concept has been introduced briefly and a high-speed serial on-chip transceiver architecture has been shown.
It has been proposed to employ a generic core wrapper architecture which encapsulates the functional core into its common infrastructure environment for clocking, power management and NoC communication. From this, basic architectures and constraints for the MPSoC clock generation circuits that are developed in this work have been derived. This mainly includes requirements for small chip area, low power consumption, high maximum output frequency, fast lock-in and frequency change time. The circuit design and implementation flow being used to realize clock generator circuit in deep-submicron CMOS technologies has been summarized. Finally, various testchip realizations, containing the components developed in this work have been presented initially. They will be referenced in the following sections where measurement results of the clock generators are shown.

3 Digitally Controlled Oscillators

The realization of on-chip clock generators requires controllable oscillator circuits for frequency generation. This chapter first gives an overview over state-of-the-art DCO topologies and tuning mechanisms. Based on this, novel multi-phase DCO circuit realizations in both 65nm and 28nm CMOS technology are presented, including new concepts for robustness with respect to PVT variations. Differential clock buffers distributing the multi-phase clock signals are analyzed in detail with special focus on their phase error correction properties, which is essential for implementations in deep-submicron CMOS technologies with severe process variations.

3.1 Overview

Controlled oscillators are the key components of PLL based clock generators. They generate a clock, which frequency is adjustable by a tuning signal. In PLLs with analog loop filters, the oscillator tuning signal is analog as well, e.g. a voltage (voltage controlled oscillator (VCO)) or a current (current-controlled oscillator (CCO)). The DCOs considered in this work are tunable by a digital tuning word for applications in ADPLLs as shown in Sec. 4.1. Usually DCOs require more chip area compared to their analog counterparts [TRF08], but the purely digital control scheme benefits from technology scaling of digital gates in advanced CMOS technologies and provides advantages with respect to functionality and design implementation (see Sec. 4.1).

For the targeted on-chip clocking applications in this work, *ring oscillators* are considered as DCO topologies. Although LC oscillators provide low jitter clocks, which are mandatory for RF transceiver applications for example, they require integrated inductors with extremely large chip area and are therefore not suitable for local clocking applications within heterogeneous MPSoCs where multiple clock generator instances are required on one chip.

In general, a tunable ring oscillator consists of a chain of controllable delay elements which are connected in a feedback loop. In the single-ended version as shown in Fig. 3.1(a) an *odd* number of inverting elements is required to satisfy the oscillation

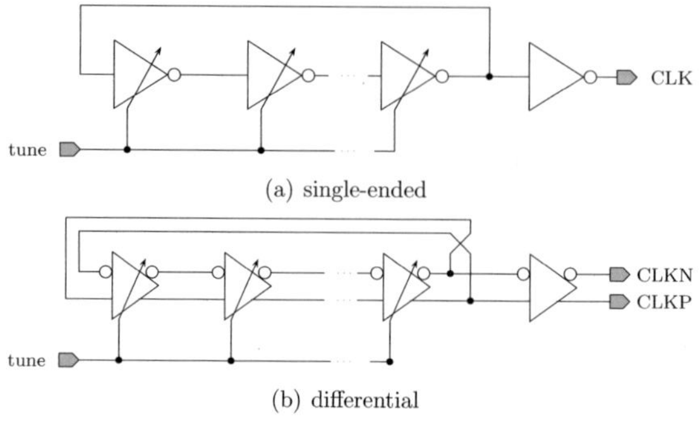

Figure 3.1: Ring oscillator topologies

criterion [MR09], [Bak05]. In the differential case as shown in Fig. 3.1(b) the number of delay elements can be *even* when an additional 180° phase shift is introduced by twisting of differential signal lines. Commonly the clock outputs are buffered to decouple the internal oscillator ring from the external load capacitance that might vary in a wide range. Multiple of the internal oscillator signals can be fed to output buffers to generate clock signals with the same frequency but defined shifts in phase. These topologies are referred to as *multi-phase* oscillators.

Tuning is achieved by controlling the delay of the active stages within the ring. Individual tuning mechanisms allow to adjust the delay per-stage. In contrast, centralized tuning circuits allow common delay adjustment of all stages in the ring. The digitally controllable delay cells as used in DCOs are also applicable for DLL based clock generators (see. Sec. 2.6) or for timing adjustments in data transmission circuits, as for example serial high-speed NoC links as shown in Sec. 2.4.

The DCO produces a clock signal which depends on the digital tuning signal c and the PVT conditions of the chip

$$T_{\mathrm{DCO}} = F(\mathrm{PVT}, c) \tag{3.1}$$

where F is a generally non-linear function depending on the circuit topology. The basic circuit requirement is that by tuning the DCO within a closed loop ADPLL a target period of T_0 must be realizable over all PVT conditions. The main performances of DCOs for the on-chip clocking applications in this work are introduced in the following.

Tuning range A large tuning range $|T_{\mathrm{DCO}}(c_{\max}) - T_{\mathrm{DCO}}(c_{\min})|$ is required to satisfy the condition $T_{\mathrm{DCO,min}} < T_0 < T_{\mathrm{DCO,max}}$ over all specified PVT variations. This in general leads to the demand for a wide range of the digital tuning signal c, which significantly impacts the complexity of the tuning circuits and digital control.

Tuning step size The digital tuning mechanism causes a minimum tuning step size $T_{\mathrm{step}} = K_{\mathrm{t}}(c) = T_{\mathrm{DCO}}(c+1) - T_{\mathrm{DCO}}(c)$ corresponding to the least significant bit (LSB) switching of the digital tuning signal c. If this is too large, the period jitter performance is degraded. This results in the demand for a fine resolution of the tuning word c, leading to more effort in the tuning circuits as well.

Tuning linearity The tuning gain $K_{\mathrm{t}}(c) = \Delta T_{\mathrm{DCO}}/\Delta c$ effects the stability of the closed loop ADPLL system (see. Sec.4.1), controlling the DCO. Therefore the variation of $K_{\mathrm{t}}(c)$ over the tuning range and with respect to PVT variations should be as small as possible and the tuning characteristic should be monotonic.

Power consumption When oscillating, the DCO draws dynamic power from the supply net. To realize energy efficient solutions the power consumption should be as small as possible. However, there exist a fundamental trade-off between power consumption and jitter due to internal noise sources [MR09], [GKGN09], which must be considered when optimizing a low-power DCO.

Jitter The accuracy of the DCO output clock timing is degraded by various types of jitter[1] [MR09]. One main jitter source is internal noise of the devices in the DCO. Based on the application different types of jitter can be of interest. As the main target application of the DCOs in this work is core clock signal generation, the *period jitter* is considered as important performance metric. The accumulation of DCO jitter is attenuated by the closed loop ADPLL as analyzed in Sec. 4.1.

Supply noise sensitivity The sensitivity of the DCO period with respect to the supply voltage $\Delta T_{\mathrm{DCO}}/\Delta V_{\mathrm{DD}}$ should be minimized to prevent noise coupling from the power supply rails to the oscillation loop. This is especially important, when multiple oscillators are powered by the same supply net within the MPSoC. Although a dedicated low noise supply net (e.g. from a linear regulator) can be used for only the DCOs, coupling between them can increase clock jitter.

[1] For details see App. A.1.

Area For per-core instantiation of clock generators within MPSoCs the chip area must be minimized to reduce the area overhead of individual clocking. With respect to the DCO architecture a minimization of the width of the digital tuning word c can significantly reduce chip area, both in the DCO and its ADPLL control circuits, which is in contrast to the requirements for tuning range and tuning step size.

3.1.1 Digital Tuning Mechanisms for Ring Oscillators

The tuning circuit topology mainly effects the performances of the DCO as described in the previous subsection. In the following different state-of-the-art tuning methods are presented.

3.1.1.1 Chain Length Adjustment

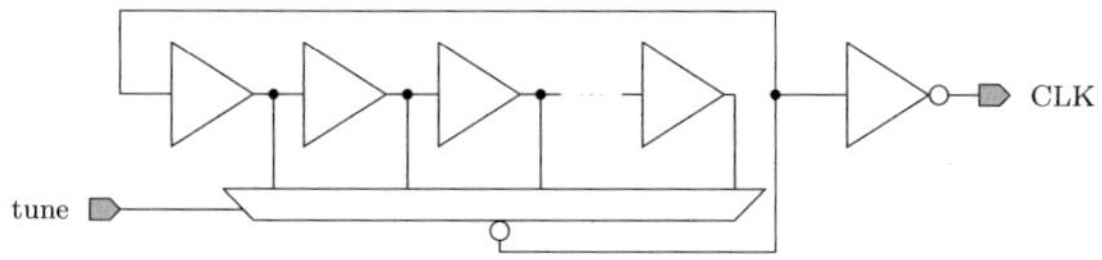

Figure 3.2: DCO with selectable chain length

The oscillation period of a ring oscillator depends on the number n of delays in the oscillation loop. By insertion of multiplexers this number can be adjusted for tuning as shown in Fig. 3.2. The period reads

$$T_{\mathrm{DCO}} = 2 \cdot (t_0 + n \cdot t_{\mathrm{d}}) \tag{3.2}$$

where t_0 is the offset delay introduced by the ring length selection devices which limits the the minimum oscillation period. This topology allows the realization of a wide tuning range by adding more delay elements to the ring, but the tuning step size is limited to the double delay of a single element $2 \cdot t_{\mathrm{d}}$. This is usually in the range of tens to hundreds of picoseconds, depending on the target technology.
To overcome this issue, multiplexed delay paths can be applied as tuning elements, where the signal path is selectable between different gate chains which differ in delay, but with a step size smaller than one gate delay. Tuning step sizes down to the ps range have been reported [WWWW05], [Wag09]. But the additional multiplexing delay offset limits the minimum oscillation period as well.
No multi-phase clock signals can be generated with these topologies. The implementation of DCOs based on selectable chain lengths is efficient, due to the fact

that all required circuit components can be realized by digital standard cells. Standard digital synthesis and place and route implementation flows can be used here [EMH+07]. The trade-off between jitter due to intrinsic noise and power consumption can be directly addressed by adjustment of the drive strength of the digital delay cells. The supply noise sensitivity is large due to the fact that the gates are directly connected to the supply rails which influences their delay.
As example, DCOs with tuning by selectable delay chain lengths have been reported in [WWWW05, LJK+05, SCL07, EMH+09, Wag09, SLH+10].

3.1.1.2 Switchable Load Capacitances

Significantly finer tuning steps can be achieved by adjustment of the delay elements itself. In general the delay of a CMOS inverter stage can be represented by an intrinsic part and an output load dependent part [SN90], reading

$$t_{\mathrm{d}} = t_{\mathrm{d,intrinsic}} + \frac{C_{\mathrm{L}} \cdot V_{\mathrm{DD}}}{I_{\mathrm{D0}}} \tag{3.3}$$

where I_{D0} is the on-current of the MOS device, denoting the drain current I_{D} under the condition $V_{\mathrm{GS}} = V_{\mathrm{DS}} = V_{\mathrm{DD}}$. For simplicity a balanced delay cell is assumed, where the rising edge delays are equal to the falling edge delays. It is

$$I_{\mathrm{D0}} \propto (V_{\mathrm{DD}} - V_{\mathrm{th}})^{\alpha} \tag{3.4}$$

using the alpha-power law model from [SN90] with $\alpha < 2$ for short channel devices. The intrinsic delay of the inverter is proportional to the transition time t_{tt} at its input $t_{\mathrm{d,intrinsic}} \propto t_{\mathrm{tt}}$. Fine tuning can be achieved by changing the load capacitance C_{L} of the stage as shown in Fig. 3.3, thereby altering the second term of Eq. 3.3 which mainly contributes to the total delay. This has the advantage, that the tuning method directly offects the main delay contributor and thus reduces the minimum delay of the tuned cells.

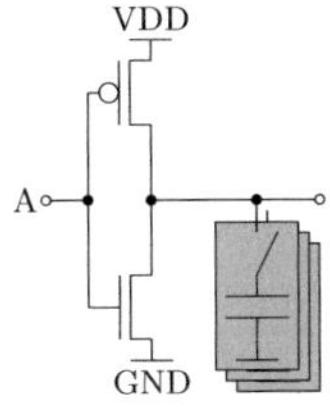

Figure 3.3: DCO stage with switched load capacitances

Digital control can be realized by using multiple capacitors which are switched by the control signal bits. The achievable tuning range is directly effected by the capacitance control range $C_{\mathrm{on}}/C_{\mathrm{off}}$ of the loads, whereas for high speed operation a low absolute capacitance value is desired according to Eq. 3.3.

Several digital capacitance control schemes have been reported previously. When using varactors the capacitance tuning range is relatively low due to the voltage insensitive part of the MOS capacitance, especially when bulk CMOS technologies are used. It is typically $C_{\mathrm{on}}/C_{\mathrm{off}} < 2$. [CCL05] proposed to use the input state dependent input capacitance of logic gates for delay tuning. Thereby capacitively tuned delay cells can be implemented using digital standard cells.

A wider capacitance control range can be achieved by using switches (e.g. CMOS transfer gates) to connect fixed load capacitors to the output node of the delay stage. This promises a wider $C_{\mathrm{on}}/C_{\mathrm{off}}$ ratio compared to the simple varactor. Additionally a large portion of the on capacitance is contributed by the load capacitance of the transfer gate itself, acting as varactor.

However, when switched on, the CMOS transfer gate shows an series resistance which leads to a lower *effective capacitance* seen by the delay stage. An estimation of this effective capacitance can be found in [QPP94]. The idea of adjusting the oscillation frequency by the switch resistance of the capacitive load has been presented in [MX00, ABR$^+$99].

[SCL07] presents hysteresis based delay cells which act as capacitive load until its switching threshold is reached and afterwards actively contribute to drive the signal transition on the delay stage output node.

As shown in Eq. 3.3 the delay stage input transition time effects its delay. In a chain of delay stages, the the signal transition time is increased when the load capacitances increase. This effect must be considered when designing a delay cell with linear digital tuning behavior. Although the second term of Eq. 3.3 suggests that a binary weighted set of capacitors together with a binary tuning signal c lead to linear characteristics, the presence of the transition time effect requires reduction of the larger capacitors to linearize the tuning characteristics. As example in the fine tuning stage in [Wag09], the load capacitance weights are (1 2 4 7 14 29) (binary would be (1 2 4 8 16 32) to linearize the tuning characteristics.

Capacitive tuning is a individual tuning method being applied per cell. This has the drawback, that for multi-phase DCOs the tuning circuits (e.g. capacitor and switch arrays) must be replicated for each stage at the cost of chip area. The intrinsic noise performance of capacitively tuned delay stages is generally good [MR09], because capacitors do not add noise to the signal. The supply voltage sensitivity is similar

to the all-digital DCOs with switched chain lengths.
As example, capacitive tuning of DCOs has been reported in [CCL05, Wag09, MX00, ARK07, ABR+99].
Close to the application context of this work, the advantages of capacitive tuning are used in the delay cells for 90° clock phase shift within the high speed serial NoC link as presented in Sec. 2.4.

3.1.1.3 Drive Strength Adjustment

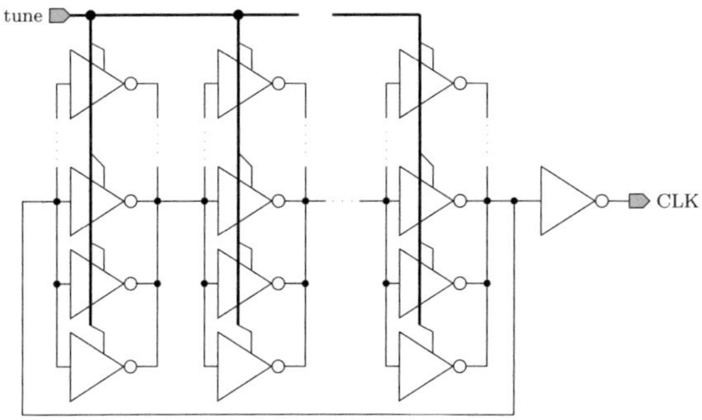

Figure 3.4: DCO with tristate inverter array

Considering Eq. 3.3 tuning can also be achieved by adjusting the drivestrenght of the delay stage in terms of I_{D0}. This can effectively be realized by parallel connection of tristate inverters operating on the same signal net. These inverter array DCOs [ON04, TRF08] consist of n stages, where for single-ended implementations n must be odd. Each stage consists of $N_{drv,max}$ parallel tristate inverters. Each of the tristate inverters form a capacitive load for the previous stage, thereby effecting the second term of Eq. 3.3. If all n stages have the same number of activated inverters, as it would be required for generation of multi-phase clock outputs with equal phase shift, the total oscillation period T_{DCO} reads

$$T_{DCO} = 2 \cdot n \cdot \left(\frac{C_L \cdot V_{DD}}{I_{D0}} \cdot \frac{N_{drv,max}}{N_{drv,on}} + t_{d,intrinsic} \right) \tag{3.5}$$

where $N_{drv,on}$ is the number of activated tristate inverters. The tuning step size can be approximated by

$$T_{step} = \left| \frac{dT}{dN_{drv,on}} \right| = 2 \cdot n \cdot \frac{C_L \cdot V_{DD} \cdot N_{drv,max}}{I_{D0}} \frac{1}{N_{drv,on}^2} \tag{3.6}$$

The tuning range corresponds to the maximum and minimum numbers of inverters that are on, i.e. the fill factor of the inverter array. The lower bound of the tuning range is limited by the maximum number of on inverters

$$T_{\mathrm{DCO,min}} = 2 \cdot n \cdot \left(\frac{C_{\mathrm{L}} \cdot V_{\mathrm{DD}}}{I_{\mathrm{D0}}} + t_{\mathrm{d,intrinsic}} \right) \tag{3.7}$$

The upper bound of the tuning range given by the minimum number of inverters that must be on, for a maximum allowed period tuning step T_{step}, which is given Eq. 3.6. Neglecting the intrinsic delay of the inverters $t_{\mathrm{d,intrinsic}}$ at this large period tuning point, this results in

$$T_{\mathrm{DCO,max}} = \sqrt{2 \cdot n \cdot \frac{C_{\mathrm{L}} \cdot V_{\mathrm{DD}}}{I_{\mathrm{D0}}} \cdot N_{\mathrm{drv,max}} \cdot T_{\mathrm{step}}}. \tag{3.8}$$

Therefore the tuning range for a given step size can be increased by increasing the total number of tristate inverters in the array. As example, Fig. 3.5 visualizes the analysis results of a 3 stage ring DCO in 65nm CMOS technology with tristate inverter tuning with $N_{\mathrm{drv,max}}$ tuning cells per stage. It shows the trade-off between achievable tuning range at a given maximum tuning step in Fig. 3.5(c), where better performance can be achieved by increased $N_{\mathrm{drv,max}}$. For reasonable tuning ranges to cover PVT variations and achieving a small tuning step in the ps range, hundreds of inverters are required [TRF08], which leads to large chip area and high power consumption. Although this reduces jitter due to device noise, this tuning scheme is not suited for compact DCOs for per-core instantiation. The decoding logic for the thermometer coded array tuning consumes additional area [TRF08].
As example, inverter array DCOs have been reported in [ON04, ARK07, TRF08, ZAJ+11]. The drive strength of the delay cells can also be adjusted directly by insertion of switched resistors in series to the driving devices in the delay stages, as presented in [ZK08, LJK+05, SKK08, HWES10].

3.1.1.4 Current-starved Inverters

Current-starved inverter delay cells as shown in Fig. 3.6 use current source devices which directly define the maximum output current of the inverter stage and thereby its output delay represented by the second term of Eq. 3.3 [MR09]. The tuning current is generated by a DAC with current output, which is then mirrored to all delay stages of the DCO. This centralized tuning scheme reduces the area overhead, especially for multi-phase DCOs with common tuning for each stage. A wide tuning range can be achieved with this topology. The current source devices

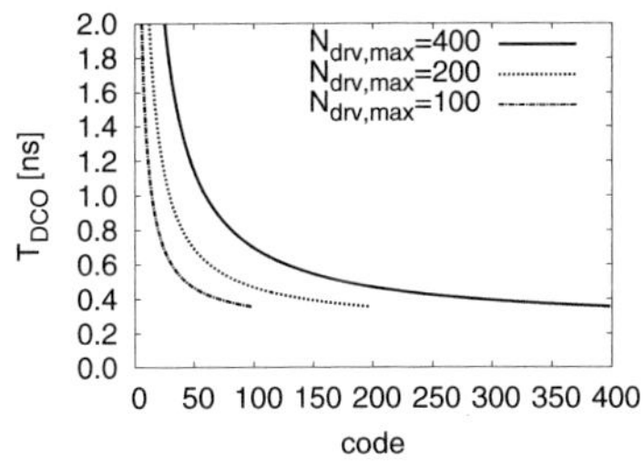

(a) tuning characteristics

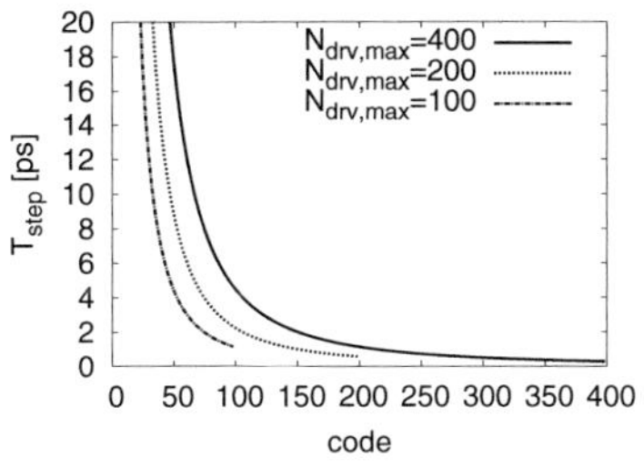

(b) tuning step size

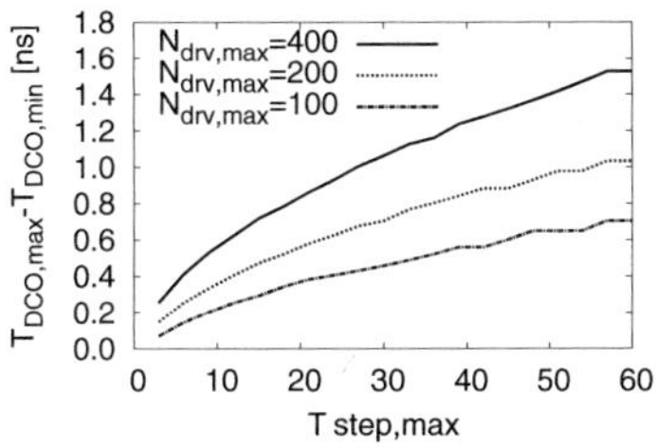

(c) tuning range and step size trade-off

Figure 3.5: Inverter array DCO analysis results

in each stage having high output resistance significantly reduce the supply noise sensitivity. However, the DAC and its bias circuitry generate noise which is coupled to the oscillation loop and results in jitter.

Obviously, also voltage tuned delay cells as parts of VCOs can be used as DCO core ring oscillators, when its tuning voltage is provided by a DAC. But as most of the commonly used VCO delay cells employ the cell current as real tuning property and use a voltage-to-current converter, this can be simplified by directly realizing a DAC with current output.

Digitally controlled delay elements using current-starved inverters have been reported in [SLM01, MNS03].

3.1.1.5 Supply Voltage Regulation

As shown in Eq. 3.3 and Eq. 3.4 the delay of a CMOS stage significantly depends on the supply voltage V_{DD}, which therefore can be used for DCO tuning. Besides the direct application of a tuning supply voltage from a digitally controllable voltage regulator, the addition of a controllable resistance in the supply net of the DCO as

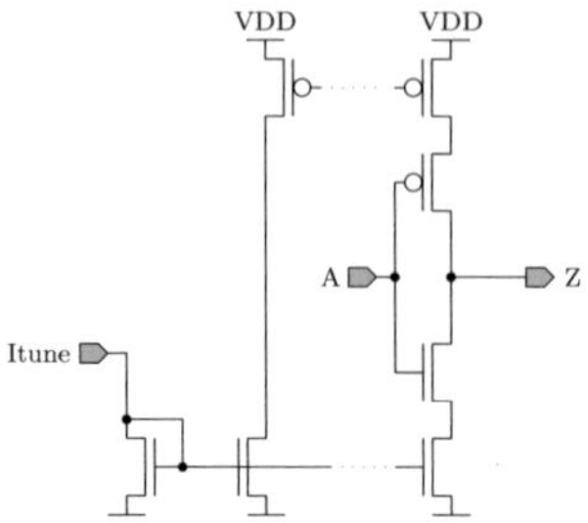

Figure 3.6: Current starved inverter delay cell

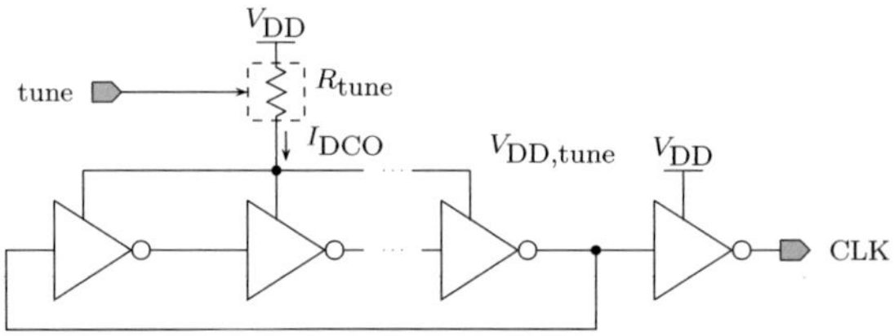

Figure 3.7: DCO with supply voltage tuning

shown in Fig. 3.7 serves the same purpose. The current consumption of the logic causes a voltage drop over the tuning resistor

$$V_{DD,tune} = V_{DD} - I_{DCO} \cdot R_{tune}. \tag{3.9}$$

The current consumption of the core can be estimated to be proportional to the oscillation frequency $1/T_{DCO}$ and the tuning supply voltage

$$I_{DCO} = \frac{C' \cdot V_{DD,tune}}{T_{DCO}} \tag{3.10}$$

where C' is the effective capacitance of the ring oscillator core. From this it can be concluded that

$$T_{DCO} = \frac{C' \cdot V_{DD,tune} \cdot R_{tune}}{V_{DD} - V_{DD,tune}}. \tag{3.11}$$

Using in Eq. 3.3 and Eq. 3.4 the oscillation period dependency from the tuning supply voltage can be expressed by

$$T_{\mathrm{DCO}} \propto \frac{C_{\mathrm{L}} \cdot V_{\mathrm{DD,tune}}}{(V_{\mathrm{DD,tune}} - V_{\mathrm{th}})^{\alpha}}. \tag{3.12}$$

At the target tuning point Eq. 3.11 and Eq. 3.12 must be fulfilled. Fig. 3.8(a) shows the tuning characteristics of a DCO with tuning resistance in the supply net based on numerical evaluation of Eq. 3.11 and Eq. 3.12 for different values of C', representing different gate sizes of the oscillator core ring. For this analysis it is $V_{\mathrm{th}} = 0.45\mathrm{V}$, $V_{\mathrm{DD}} = 1.0\mathrm{V}$ and $T_{\mathrm{DCO}} = 0.1 \cdot 10^{-9} \cdot \mathrm{Vs} \cdot V_{\mathrm{DD,tune}}/(V_{\mathrm{DD,tune}} - V_{\mathrm{th}})^2$. As result, almost linear tuning characteristics of T_{DCO} over R_{tune} can be achieved.

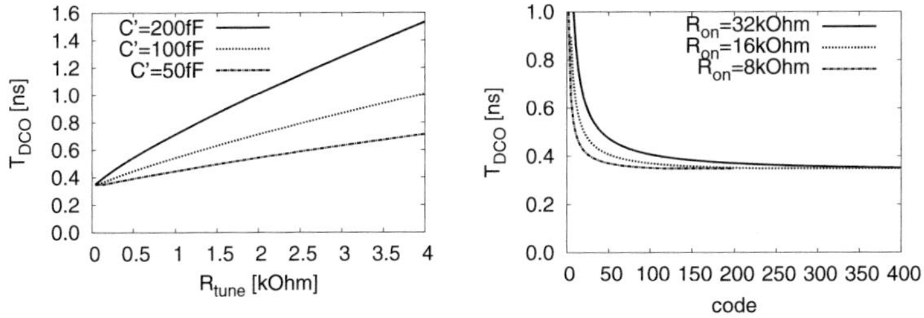

(a) resistance tuning

(b) digitally tuned resistor

Figure 3.8: DCO with supply resistance tuning model analysis results

When realizing the controlled resistor by parallel connection of MOS transistor switches each with R_{on}, where c are on, the tuning resistance reads

$$R_{\mathrm{tune}} = \frac{R_{\mathrm{on}}}{c}. \tag{3.13}$$

Fig. 3.8(b) shows the resulting tuning characteristics for different base R_{on} values (for $C' = 100\mathrm{fF}$). Generally a higher number of MOS switches with higher on resistance R_{on} lead to a wider tuning range with smaller step size. This tuning behavior is similar to the inverter array DCOs presented in Sec. 3.1.1.3. Here also hundreds of switches are required for acceptable tuning ranges with fine tuning steps. But here the supply path resistors are applied in a centralized tuning scheme where the switches are *shared* among all DCO stages. This significantly reduces the area overhead and makes this tuning mechanism feasible for compact DCO implementations. Details on the circuit implementation of a DCO with resistive supply voltage tuning can be found in Sec. 3.3 of this work. The supply voltage tuning scheme can provide a wide tuning range with small step sizes. One drawback is that the resistive elements in the supply path add noise to the DCO. This can be prevented to some extend by adding decoupling capacitors to the $V_{\mathrm{DD,tune}}$ net. DCOs with supply voltage tuning are commonly used where multi-phase outputs are required, like for example for Flying Adder frequency synthesis as presented in [Xiu07, XLL12]. Parallel connected PMOS devices as resistance in the supply path are used in [KSK+09].

3.1.2 Combining Tuning Mechanisms

Each of the different tuning methods has individual benefits with respect to the main DCO requirements. In order to cope with the trade-off between tuning range, tuning step size and circuit complexity of individual tuning methods, multiple of them can be combined within one DCO circuit [SCL07], [WWWW05], [ZK08], [LJK+05], [SLH+10], [YCYL12]. Basically different types of controlled delay elements are connected in series within the ring oscillator loop. Coarse tuning elements are employed for calibration purposes with respect to process variations. Fine tuning is then commonly used for phase and frequency tracking during operation. The benefit of this is the reduced gain of the DCO characteristics, which leads to less jitter introduced by the control signal LSB during closed loop ADPLL operation. As example a switched delay chain can be used for coarse adjustment of the DCO period whereas fine tuning is achieved by delay cells with digitally adjustable capacitive loads and others with drive strength adjustment [ZK08].

Therefore the digital tuning word c is split into sub-words, each controlling a different tuning mechanism. Since they are typically realized by completely different circuit techniques their individual characteristics do no automatically match. This issue is visualized in Fig. 3.9(a), where the fine tuning range is significantly larger than the LSB step of the coarse tuning. When operating the DCO at such a point of unwanted non-linearity, large tuning induced period jitter can be generated. These non-linearities are strongly effected by variations (e.g. device mismatch) in the manufactured circuits.

Basically there are two solutions to this issue. The DCO characteristics can be calibrated such that overflow values are defined for the fine tuning words, which correspond to one LSB step of coarse tuning, as shown in Fig. 3.9(b). Circuits that allow this kind of self calibration have been presented in [CYL09, Wag09]. They require significant additional circuit overhead which increases the complexity of the whole clock generator.

As new alternative approach, the 65nm DCO developed in this work as presneted in in Sec. 3.2 also uses separated coarse and fine tuning mechanisms. But by a new compensation technique for the supply voltage and temperature dependency of the DCO period it is ensured that after applying the coarse tuning for process calibration, operation can be achieved only with fine tuning within the specified ranges for supply voltage and temperature.

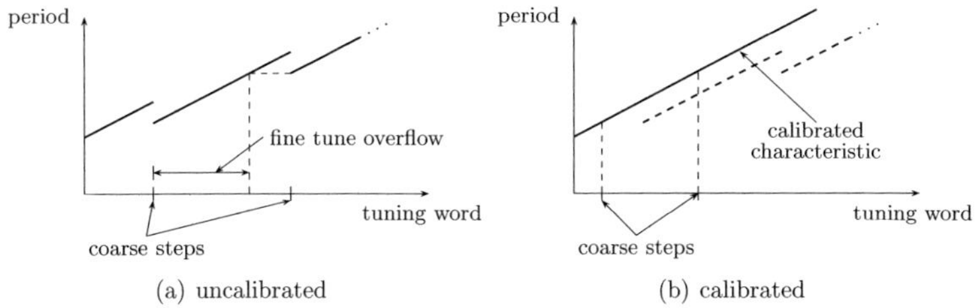

Figure 3.9: DCO characteristics with multiple tuning mechanisms

3.2 A Multi-phase DCO in 65nm CMOS Technology

For the targeted application within an ADPLL clock generator for multi-phase output clocks, a new DCO based on a fully differential ring oscillator topology is developed. Tuning is achieved by current starved inverters that are controlled by a current-based DAC. This circuit has been implemented in 65nm CMOS technology in the "Tommy" testchip and, with some modifications in the tuning and bias circuitry, in the "Atlas" testchip, as presented in Sec. 2.8. Its detailed circuit structure is presented in the following subsections.

3.2.1 Circuit Structure

Fig. 3.10 shows the schematic of the DCO ring oscillator [HEH+13]. It is built up using four pseudo-differential stages with current-starved inverter cells as shown in Fig. 3.11(a). The tuning bias signals tp and tn are generated by a current-based DAC, and are applied to all DCO stages. This centralized tuning approach ensures symmetry of the multi-phase output signal. Within the pseudo-differential stages, 180° phase shift between the differential nodes is ensured by cross-coupled inverters. This positive feedback additionally compensates delay mismatch variations in the DCO cells. The output signal of each cell is buffered by a conventional CMOS inverter. The four differential stages are connected with an inverted feedback to provide an additional 180° phase shift to fulfill the oscillation criterion. This topology provides eight equally spaced output clock phases with 45° phase shift, corresponding to a relative delay of $T_0/8$ when the DCO is locked to a period of T_0.

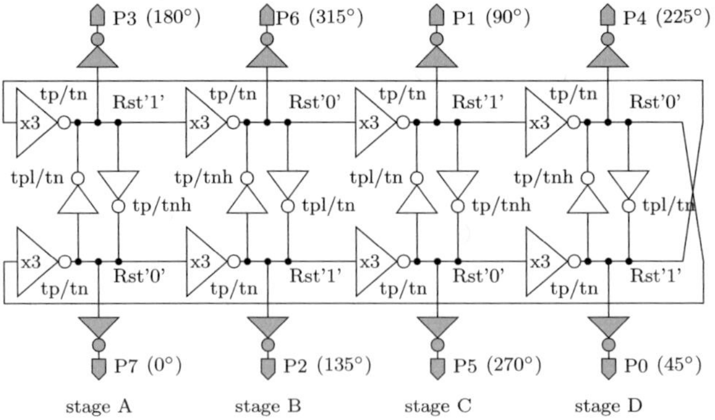

Figure 3.10: DCO core schematic, [HEH+13]

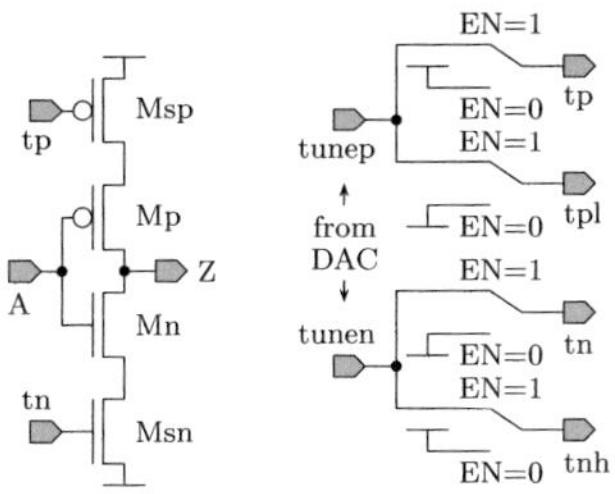

(a) cell (b) tune switch

Figure 3.11: Current-starved inverter DCO tuning schematics, [HEH+13]

As presented in [HEH+13], the ring oscillator with an even number of differential stages might suffer from startup problems because a stable common mode can be reached, when the differential gain of the delay stages is too small to meet the oscillation criterion. In this case the differential nodes of each oscillator stage settle on the same voltage level, where the following differential stage has the inverted common mode. This common mode does not lead to an oscillation because an even number (4) of differential inverters is in the loop. This issue can be overcome by higher drive strengths of the cross-coupled inverters that ensure differential signals at the according nodes. If they are too strong, the main oscillation loop can be slowed down and regenerative switching can occur which leads to increased jitter [MR09]. If they are to weak, startup might fail. Here the drive strength of the cross-coupled inverters is chosen one third of the main inverters to prevent regenerative switching [MR09]. The differential oscillation startup issue is solved by a special power up technique as explained in the following.

When a current-starved DCO is disabled (EN=0), the main inverters are disabled. Their output nodes would be high-resistive. Therefore, the internal node voltages would not be well defined which might lead to unwanted static currents in the output inverters and startup might fail when a stable common mode is established as explained before. Fig. 3.12 shows the simulated differential small-signal DC gain of one pseudo-differential DCO stage. If the common mode is below 0.2V or above 0.7V the differential mode is attenuated. Even if differential perturbations (e.g. due to noise) are present, no differential oscillation can ramp up.

The bias voltages of the current-starved inverters are switched to dedicated levels during power down by the tuning voltage switch shown in Fig. 3.11(b). It is implemented using simple CMOS transmission gates. When the DCO is in power down mode (EN=0), the main inverters are completely disabled whereas in the

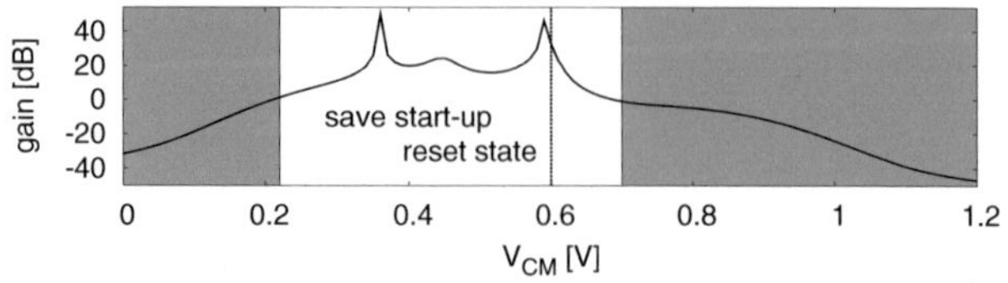

Figure 3.12: Simulated differential small-signal DC gain of one DCO stage versus common mode voltage, $V_{\mathrm{DD}} = 1.2V$, 65nm CMOS

cross-coupled stages either the pull-up current source device Msp or the pull-down current source device Msn are *enabled.* Thereby the internal differential nodes are kept in fully settled differential mode, as indicated by the Rst'1' and Rst'0' labels in Fig. 3.10. The output inverters are switched completely and no static current (except leakage) flows in power down mode. From this differential power down mode safe oscillation start-up occurs, because the DCO stages have their maximum differential gain at this reset point ($V_{\mathrm{CM}} \approx V_{\mathrm{DD}}/2$). During DCO operation (EN=1) all current source devices Msp/Msn in the current-starved inverters are connected to the tuning voltages tunep and tunen, respectively. Fig. 3.13 shows the waveform of the DCO startup. The internal nodes P0Q to P7Q are plotted together with the tuning voltages of the current-starved inverters.

A current-based DAC is used for tuning. It converts the digital tuning signals to the biasing voltages of the PMOS and NMOS sources in the current-starved inverter cells. Fig. 3.14 shows its schematic. First, a reference bias current at the input iref is multiplied by the coarse tune current switch bank. This consists of 6-bit binary weighted current switches being controlled by a the tuning signal c_{coarse}. Due to the fact that coarse tuning is only performed during start-up of the ADPLL (see. Sec. 3.2.3) the use of binary weighted switches is feasible, although they can add significant noise to the DCO by switching large transistors, especially when changing the MSB of c_{coarse}. On the other hand the binary weighted tuning signal can directly be applied from the ADPLL controller without additional decoding effort. Second, the reference current is fed to the fine tuning gain stage where it can be multiplied by a 2-bit, binary weighted control signal c_{ftgain} to set the gain of the following fine tuning stage. This allows to adjust the trade-off between the DCO tuning step size T_{step} and the period range which is covered by fine tuning. The fine tuning stage consists of 64 minimum sized switches that are controlled by a thermometer coded tuning signal c_{fine}. Thereby the control signal related jitter during operation is minimized, because only little parasitic charge is injected to the

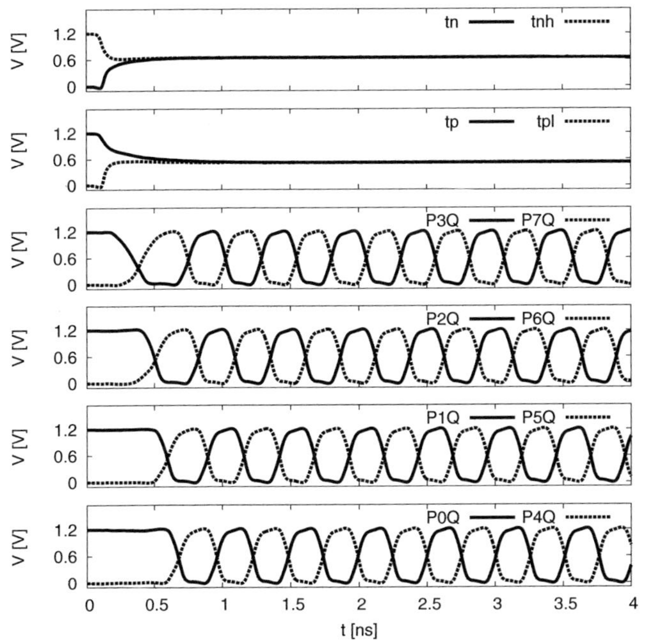

Figure 3.13: DCO startup waveform simulation result, $V_{\mathrm{DD}} = 1.2\mathrm{V}$, typical process (TT), $\theta = 27°\mathrm{C}$, $T_{\mathrm{DCO}} = 507\mathrm{ps}$

control signal nodes. However, a thermometer decoding logic is required for c_{fine} in the ADPLL controller. The output currents of the coarse and fine tuning banks are summed on the input of the DAC output current mirror and are converted into the bias voltages tunen and tunep for the current-starved inverter cells. In summary, the tuning current can be expressed by

$$I_{\mathrm{tune}} \propto I_{\mathrm{ref}} \cdot (c_{\mathrm{coarse}} + \gamma \cdot c_{\mathrm{ftgain}} \cdot c_{\mathrm{fine}}) \tag{3.14}$$

where $\gamma < 1$ is a constant.

3.2.2 DCO Circuit Implementation

The proposed DCO has been implemented in 65nm LP CMOS technology. Fig. 3.15 show the layouts of the current DAC and oscillator core. A symmetric layout is required for good phase matching of the multi-phase output signal. Fig. 3.16 shows

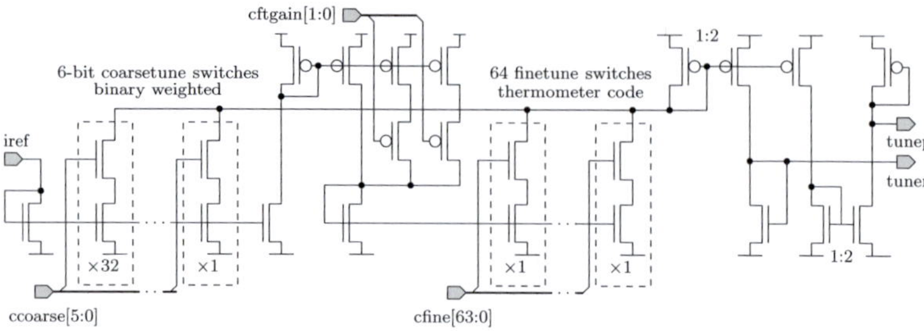

Figure 3.14: DCO tuning DAC schematic

the detailed layout of the DCO ring oscillator core consisting of four stages (A to D) as shown in the schematic Fig. 3.10. The placement sequence of D,A,C,B allows the realization of symmetric wire lengths of the internal signals in the oscillation loop. The stage outputs (P0 to P7) are connected to a symmetric 8-bit bus. Wire lengths are kept as equal as possible for all eight clock phases.

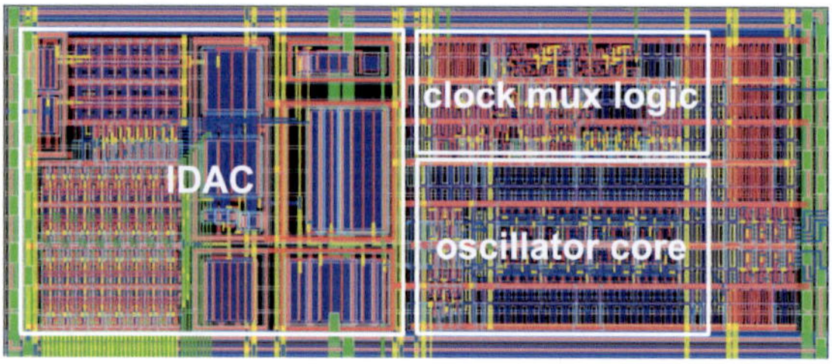

Figure 3.15: Layout of the DCO in 65nm CMOS, 24μm × 58μm

Fig. 3.17 shows the measured waveform of the free running DCO in 65nm CMOS technology. Fig. 3.18 shows the measured DCO fine tuning curves for different c_{coarse} values and fine tune gain settings. The fine tuning step size is suitably small for low jitter operation reaching from ≈ 0.36ps for $c_{\text{ftgain}} = 0$ to ≈ 1.38ps for $c_{\text{ftgain}} = 3$. Fig. 3.19 shows the measured differential nonlinearity (DNL) for the coarse and fine tune stage of the DAC respectively. The binary weighted coarse tune stage shows a maximum DNL error of ± 1.7LSB. The thermometer decoded fine tune stage shows strictly monotonic tuning behavior with DNL > -1LSB.
Tab. 3.1 summarizes the main performances of the DCO realization in 65nm CMOS technology.

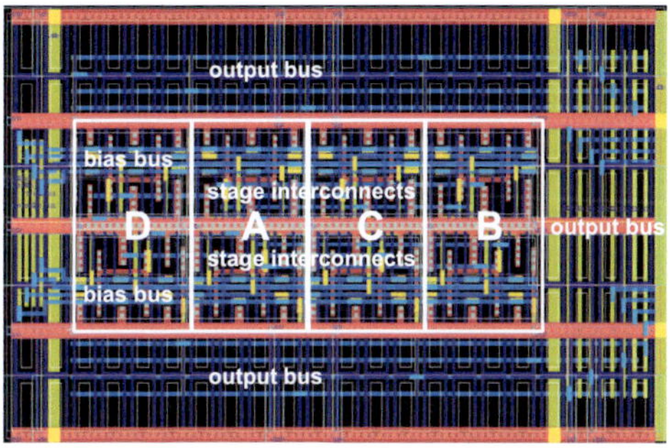

Figure 3.16: Layout of the DCO core in 65nm CMOS

Table 3.1: Typical 65nm DCO performances

T_{DCO} [ps]	500
tuning step [ps] at 2GHz	1.0
period jitter rms [ps] at 2GHz	5.4
power consumption [mW] at 2GHz	1.8

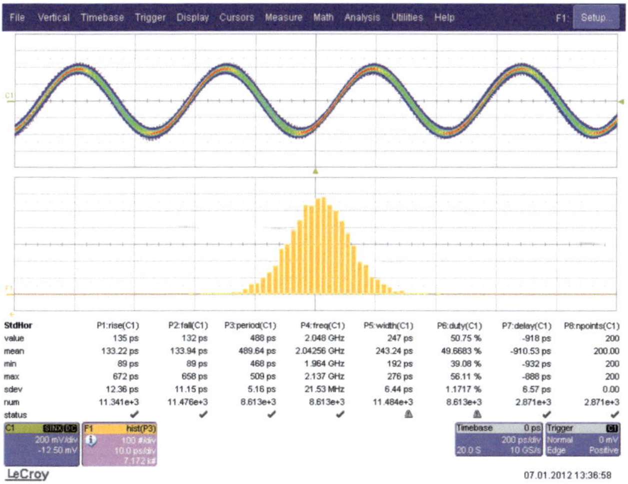

Figure 3.17: Measured 65nm DCO output waveform over LVDS pad

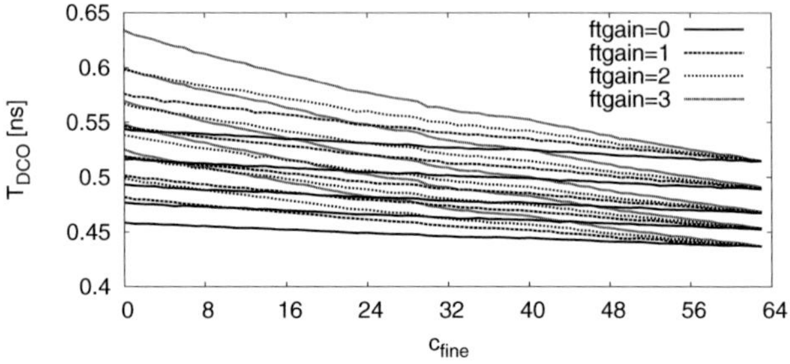

Figure 3.18: Measured DCO tuning curves at $V_{\mathrm{DD}} = 1.2\mathrm{V}$, $\theta = 25°C$, $c_{\mathrm{coarse}} = [20, 25, 30, 35, 40]$

Figure 3.19: Measured DAC DNL, 35 devices

3.2.3 ADPLL Application Scenarios

When applied in an ADPLL for MPSoC clock generation the DCO is locked to a nominal period of T_0. Fig. 3.20 illustrates a typical application scenario for DCO operation [HHH+12]. In an initial coarse tune phase the value of c_{coarse} is determined such that $T_{\text{DCO}} \approx T_0$. Linear or binary search (see Sec. 4.2.2 and Sec. 4.3.2) algorithms can be used here. During this phase the fine tune signal c_{fine} is at its middle position. During coarse lock-in mainly process variations are compensated and the DCO frequency changes in a wider range. Therefore, the output clock of the ADPLL is gated, such that the MPSoC components are not clocked during coarse lock-in. In the following fine lock phase c_{fine} is adjusted by closed-loop ADPLL operation to meet the frequency lock condition of $T_{\text{DCO}} = T_0$. It compensates the remaining coarse tune error. When the lock condition is reached, the output clock is used to run the MPSoC component(s). During operation the fine tune mechanism tracks supply voltage V_{DD} and temperature θ variations.

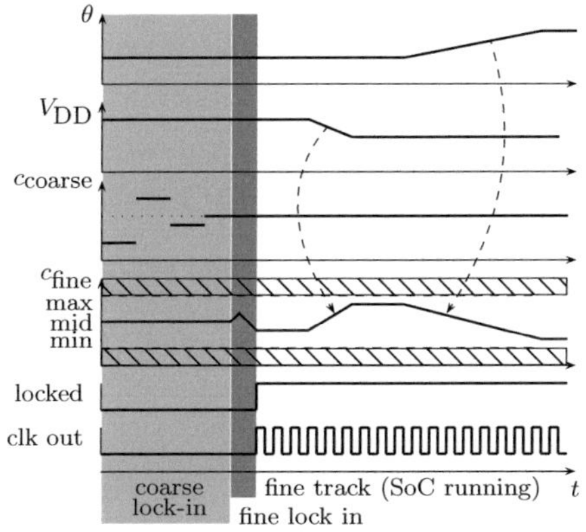

Figure 3.20: ADPLL operation phases, [HHH+12]

The MPSoC has specified operating parameter ranges for temperature $\theta_{\min} \leq \theta \leq \theta_{\max}$ and supply voltage $V_{\text{DD,min}} \leq V_{\text{DD}} \leq V_{\text{DD,max}}$. The parameters can change during system operation within these specified ranges, e.g. due to environmental temperature changes, heating by system power consumption or IR-drop in the supply networks. In order to maintain phase and frequency lock, the fine tune signal must stay within the available region $c_{\text{fine,min}} \leq c_{\text{fine}} \leq c_{\text{fine,max}}$ during system operation. To describe this phenomenon, the following assumptions are made:

- T_{DCO} has a strictly monotonic dependency from V_{DD}, for constant θ and c_{fine}.

- T_{DCO} has a strictly monotonic dependency from θ, for constant V_{DD} and c_{fine}.
- T_{DCO} has a strictly monotonic dependency from c_{fine}, for constant V_{DD} and θ, ensured by the thermometer coded fine tuning as presented in Sec. 3.2.1.
- There exists a *best-case* operating condition $(V_{\mathrm{DD}}, \theta)_{\mathrm{best}}$ where T_{DCO} reaches its minimum value for constant c_{fine}.
- There exists a *worst-case* operating condition $(V_{\mathrm{DD}}, \theta)_{\mathrm{worst}}$ where T_{DCO} reaches its maximum value for constant c_{fine}.
- Best-case and worst-case operating conditions occur at the corners of the specified operation region, where $V_{\mathrm{DD}} \in (V_{\mathrm{DD,min}}; V_{\mathrm{DD,max}})$ and $\theta \in (\theta_{\mathrm{min}}; \theta_{\mathrm{max}})$.

As shown in in Fig. 3.21 [HHH+12], the DCO period fine tune characteristics $T_{\mathrm{DCO}} = f(c_{\mathrm{fine}})$ change with (V_{DD},θ) variations in the gray regions, where the upper and lower boundaries denote the best-case $(V_{\mathrm{DD}}, \theta)_{\mathrm{best}}$ and worst-case $(V_{\mathrm{DD}}, \theta)_{\mathrm{worst}}$ corners. At initial coarse lock-in, where the fine tune signal is kept at its middle position, the actual operating condition $(V_{\mathrm{DD}}, \theta)$ is not known. However, the fine tune mechanism must be capable to compensate for all variations within the specified operating parameter range. Therefore, the criteria for safe system operation within the specified operating parameter ranges are:

- coarse lock-in at $(V_{\mathrm{DD}}, \theta)_{\mathrm{best}}$ and occurrence of $(V_{\mathrm{DD}}, \theta)_{\mathrm{worst}}$ during operation, $\rightarrow c_{\mathrm{fine}} \leq c_{\mathrm{fine,max}}$
- coarse lock-in at $(V_{\mathrm{DD}}, \theta)_{\mathrm{worst}}$ and occurrence of $(V_{\mathrm{DD}}, \theta)_{\mathrm{best}}$ during operation, $\rightarrow c_{\mathrm{fine}} \geq c_{\mathrm{fine,min}}$

as illustrated in Fig. 3.21. These constraints can be fulfilled by different approaches:

1. A wider tuning range by *more fine tuning steps* at cost of chip area.
2. A wider tuning range by larger fine tuning step size at cost of higher period jitter in the ADPLL output signal.
3. Reduced dependency of the fine tuning characteristics from (V_{DD},θ) by circuit design techniques.

For the proposed DCO implemented in 65nm CMOS technology, the third approach is chosen, whereas the 28nm DCO circuit presented in Sec. 3.3 uses the first option while benefiting from technology scaling.

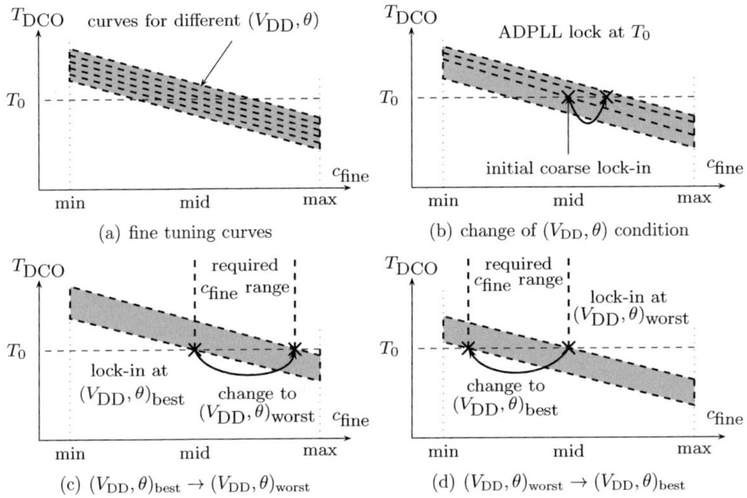

Figure 3.21: Illustration of DCO coarse lock-in at different (V_{DD}, θ) conditions and fine tune variations during system operation, [HHH+12]

3.2.4 DCO Bias Circuit

A novel biasing technique for active compensation of supply voltage and temperature variations in DCOs in order to circumvent the trade-off between PVT variation robustness and required tuning range has been developed in this work and is explained in the following subsection as it has been presented in [HHH+12].

3.2.4.1 Supply Voltage and Temperature Dependency

Although the period of oscillation of a current-starved ring oscillator is defined by the current-sources in the inverter-cells, it depends on the supply voltage V_{DD} and the temperature θ by

$$T_{\mathrm{DCO}} = F(\theta, V_{\mathrm{DD}}, I_{\mathrm{ref}}) \tag{3.15}$$

for constant values of c_{coarse} and c_{fine}, where F denotes a nonlinear function. I_{ref} is the reference current of the tuning DAC. Fig. 3.22 shows an example simulation result of the temperature and supply voltage dependency of the DCO in 65nm CMOS technology from Sec. 3.2.1 for a constant reference current I_{ref}. With increasing supply voltage the period decreases because the current source devices provide increased output current, due to their finite output resistance. Additionally, the switch devices in the current-starved inverter cells (Fig. 3.11(a)) show smaller on-resistance. With increasing temperature the threshold voltage of the switching devices decreases, leading to reduced oscillation period as well.

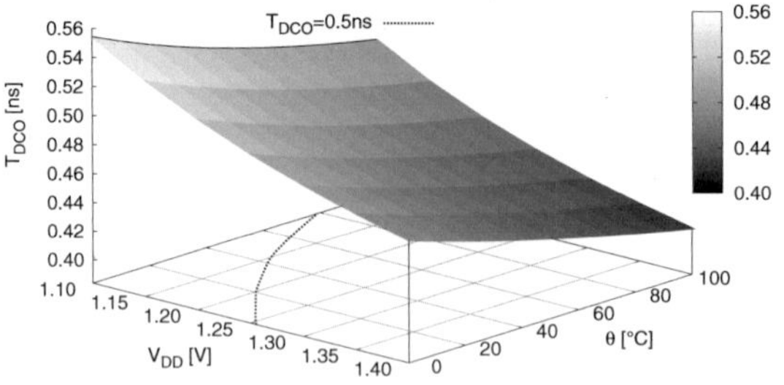

Figure 3.22: Simulated DCO period for constant $I_{\mathrm{ref}} = 30\mu\mathrm{A}$, [HHH+12]

3.2.4.2 Previous Work

Several approaches for compensation of supply voltage and temperature influences on the oscillation period of ring oscillators have been reported previously. [SAA06] and [TLC+10] compensate the ring oscillator frequency for temperature and process by adaptive biasing using threshold voltage sensing circuits. [ZA11] presents a compensation technique using an addition based current source. However this approach requires a reference gate-source voltage which can not be used for tuning because the temperature compensation is optimized for a fixed gate-source voltage. In [YYL11] an all-digital low frequency reference oscillator with PVT compensation is presented, which is based on on-chip evaluation of the relative delay of different logic gate types. This topology is not suited for high-speed ring-oscillations with current-starved tuning mechanism. [SCJ+11] presents a special DCO ring topology to reduce the supply voltage influences on the oscillation period.

The drawback of the previous work is that sensitivity versus supply voltage changes are not compensated separately from the process and temperature related effects, which is especially critical for circuits in small CMOS technologies, where short channel effects increase significantly. In contrast, this work presents selective compensation for supply and temperature related effects. Thereby the fine tune range of DCO can be reduced significantly or the fine tune step size can be decreased with the same number of control bits. This additionally can reduce the DCO gain K_t and therefore the output jitter of the ADPLL clock generator.

3.2.4.3 Bias Current Compensation Architecture

When the oscillator is locked to a specified period the (V_DD,θ) variations are compensated by the fine tuning mechanism during closed-loop ADPLL operation. As an example Fig. 3.23 shows the tuning current for a constant period of $T_0 = 0.5$ns versus temperature and supply voltage variations. The maximum required current tuning range ($\Delta I_\mathrm{ref} = I_\mathrm{ref,max} - I_\mathrm{ref,min}$) must be covered by the fine tuning stage of the DAC. If ΔI_ref is large, a large number of fine tune switches are required, which leads to larger chip area, or the fine tune step size must be increased, which leads to larger jitter.

To circumvent this trade-off, it is proposed to provide a reference current I_ref, which compensates for temperature and supply voltage variations and thereby decreases the required DAC fine tuning range. Therefore, the current bias source must show

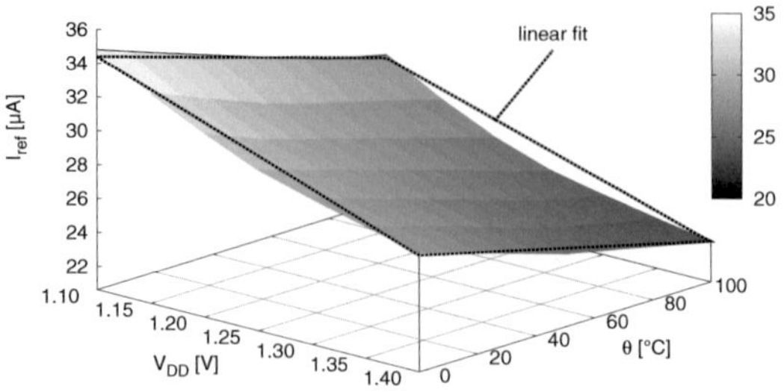

Figure 3.23: I_{ref} for $T_{\mathrm{DCO}} = 500\mathrm{ps}$, from reverse interpolation of simulation data of the DCO in 65nm CMOS, [HHH+12]

the *inverse* characteristics of the DCO core $T_{\mathrm{DCO}} = F(\theta, V_{\mathrm{DD}}, I_{\mathrm{ref}})$, which is

$$I_{\mathrm{ref}} = G_{I,\mathrm{ref}}(\theta, V_{\mathrm{DD}})|_{T_0} . \tag{3.16}$$

Obviously the function $G(\theta, V_{\mathrm{DD}})$ is nonlinear as well as shown in Fig. 3.23). It can be approximated by a two dimensional polynomial expression [Sem97]

$$I_{\mathrm{ref}} \approx c_{\mathrm{comp}} + a_{\mathrm{comp}} \cdot V_{\mathrm{DD}} + b_{\mathrm{comp}} \cdot \theta + a_2 \cdot V_{\mathrm{DD}}^2 + b_2 \cdot \theta^2 + d_2 \cdot V_{\mathrm{DD}} \cdot \theta + \ldots \tag{3.17}$$

It is proposed to neglect the higher order terms of Eq. 3.17 and to employ a current source which has linearized characteristics of $G_{I,\mathrm{ref}}$ versus θ and V_{DD}. This cancels out first-order supply voltage and temperature variation effects. The higher order error remains and is compensated by the fine tuning mechanism of the DCO. The linearized compensated reference current has *three* degrees of freedom a_{comp}, b_{comp} and c_{comp} with

$$I_{\mathrm{ref,comp}}(V_{\mathrm{DD}}, \theta) = \underbrace{(a_{\mathrm{comp}}, b_{\mathrm{comp}}, c_{\mathrm{comp}})}_{\mathbf{a}} \cdot (V_{\mathrm{DD}}, \theta, 1)^{\mathrm{T}}. \tag{3.18}$$

Therefore, the architecture of this reference current source as shown in Fig. 3.24 consists of *three* independent bias currents with different supply voltage and temperature characteristics. The first component $I_{\mathrm{ref,0}}$ is independent from θ and V_{DD}, whereas the second ($I_{\mathrm{ref,1,ptk}}$) and third ($I_{\mathrm{ref,2,pvk}}$) components show strong dependency from θ and V_{DD} respectively. The reference current for the DAC is generated by summing up these three components

$$I_{\mathrm{ref}}(V_{\mathrm{DD}},\theta) = \underbrace{(k_0,k_1,k_2)}_{\mathbf{k}} \cdot (I_{\mathrm{ref},0}, I_{\mathrm{ref},1,\mathrm{ptk}}, I_{\mathrm{ref},2,\mathrm{pvk}})^{\mathrm{T}}. \tag{3.19}$$

The weighting factors $\mathbf{k} = (k_0, k_1, k_2)$ are adjustable by a programmable current bank based on switchable current sources. Although being designed for no ($I_{\mathrm{ref},0}$) or main dependency from a single parameter θ ($I_{\mathrm{ref},1,\mathrm{ptk}}$) or V_{DD} ($I_{\mathrm{ref},2,\mathrm{pvk}}$), all three current components show parasitic dependency from V_{DD} and θ. This is considered by modeling these influences with a first-order (linear) approximation

$$\begin{pmatrix} I_{\mathrm{ref},0} \\ I_{\mathrm{ref},1,\mathrm{ptk}} \\ I_{\mathrm{ref},2,\mathrm{pvk}} \end{pmatrix} = \underbrace{\begin{pmatrix} a_0 & b_0 & c_0 \\ a_1 & b_1 & c_1 \\ a_2 & b_2 & c_2 \end{pmatrix}}_{\mathbf{A}} \cdot \begin{pmatrix} V_{\mathrm{DD}} \\ \theta \\ 1 \end{pmatrix} \tag{3.20}$$

where each bias component source has its individual set of coefficients (a, b, c). The main (wanted) components are c_0, b_1 and a_2, where all remaining components model parasitic influences. The total reference current can be expressed by

$$I_{\mathrm{ref}}(V_{\mathrm{DD}},\theta) = \mathbf{k} \cdot \mathbf{A} \cdot (V_{\mathrm{DD}}, \theta, 1)^{\mathrm{T}} \tag{3.21}$$

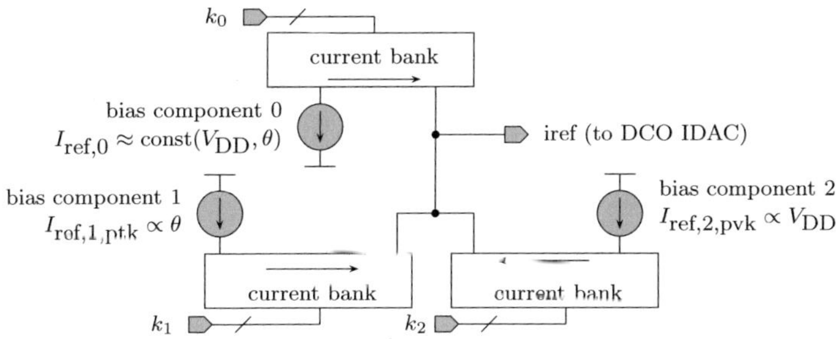

Figure 3.24: Bias current source with adjustable temperature and supply voltage dependency, [HHH+12]

The circuit realization of the three bias current components is explained in the following.

Current bias component 0 The first bias current component $I_{\mathrm{ref},0}$ has a low sensitivity with respect to θ and V_{DD}. A beta-multiplier based current reference

[Bak05] as shown in Fig. 3.25 is used here. The current I_0 is determined by M1, M2 and the resistor R. The amplifier circuit ensures that

$$V_{\mathrm{GS},1} = V_{\mathrm{GS},2} + R \cdot I_0 \tag{3.22}$$

and the equally sized devices M3 and M4 ensure that the M1 and M2 have the same current I_0. The width ratio of M2 and M1 is $W_2 = K \cdot W_1$. Assuming M1 and M2 to operate in saturation region it is

$$I_0 = \frac{\beta}{2} \cdot (V_{\mathrm{GS},1} - V_{\mathrm{th}})^2 = K \cdot \frac{\beta}{2} \cdot (V_{\mathrm{GS},2} - V_{\mathrm{th}})^2 \tag{3.23}$$

with $\beta = KP \cdot W/L$. Thereby I_0 can be expressed as

$$I_0 = \frac{2}{R^2 \cdot \beta} \cdot \left(1 - \frac{1}{\sqrt{K}}\right)^2. \tag{3.24}$$

The reference current does not depend on the supply voltage V_{DD} in a first-order approximation. The temperature dependency of I_0 can be written as

$$\frac{1}{I_0} \cdot \frac{\delta I_0}{\delta \theta} = -\frac{2}{R} \cdot \frac{\delta R}{\delta \theta} - \frac{1}{\beta} \cdot \frac{\delta \beta}{\delta \theta}. \tag{3.25}$$

It is $\delta\beta/\delta\theta < 0$ because the charge mobility is decreasing with increasing temperature. In order to achieve $\delta I_0/\delta\theta = 0$, the resistor must exhibit a postitive temperature dependency $\delta R/\delta\theta > 0$ according to Eq. 3.25. An n-well resistor is used for this purpose. The reference current $I_{\mathrm{ref},0}$ is generated from I_0 by a switchable PMOS current bank.

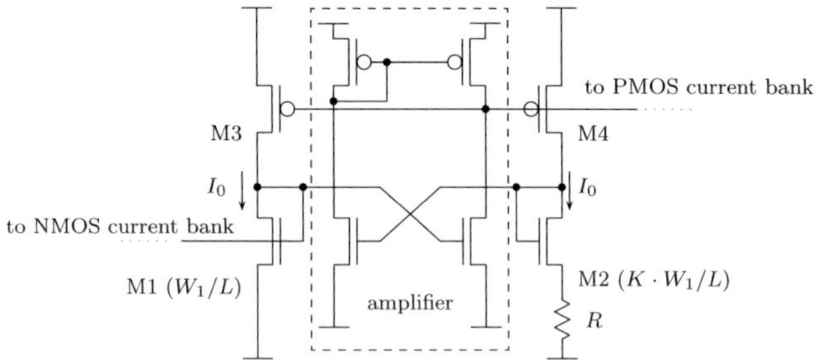

Figure 3.25: Beta-multiplier current reference for $I_{\mathrm{ref},0}$ and $I_{\mathrm{ref},1,\mathrm{ptk}}$, power-down switches and start-up circuit not shown, [HHH+12]

Current bias component 1 The second bias component $I_{\text{ref},1}$ provides a significant temperature dependency but a low supply voltage dependency. The same beta-multiplier circuit as for $I_{\text{ref},0}$ is employed, except that a poly resistor with negative temperature dependency $\delta R/\delta\theta < 0$ is used. Thus the temperature sensitivity is $\delta I_1/\delta\theta > 0$ according to Eq. 3.25. The reference current $I_{\text{ref},1}$ is generated from I_1 by a switchable NMOS current bank.

Current bias component 2 The third bias component $I_{\text{ref},2}$ shows a significant dependency on V_{DD}. Fig. 3.26 shows its schematic realization. The resistive divider R1,R2 defines a reference voltage V_{ref} which linearily depends on V_{DD}. The transistor M2 sources a current I_2 through R3, with $R_3 = R_1$. An error amplifier senses the voltage difference over R1 and R2 and adjusts the $V_{\text{GS},2}$ of M2 until $V_{\text{R3}} = V_{\text{R1}}$ and therefore the currents through R1 and R2 are equal (I_2). It is

$$I_2 = \frac{1}{R_3} \cdot \frac{R_1}{R_1 + R_2} \cdot V_{\text{DD}} = \frac{1}{R_1 + R_2} \cdot V_{\text{DD}} \tag{3.26}$$

The sensitivities with respect to V_{DD} and θ read

$$\frac{\delta I_2}{\delta V_{\text{DD}}} = \frac{1}{R_1 + R_2} \tag{3.27}$$

$$\frac{1}{I_2} \cdot \frac{\delta I_2}{\delta\theta} = -\frac{1}{R_1 + R_2} \cdot \frac{\delta(R_1 + R_2)}{\delta\theta} \tag{3.28}$$

Poly resistors are employed here, because they show a low absolute temperature dependency $\delta R/\delta\theta$, such that I_2 is mainly sensitive to V_{DD}. The reference current $I_{\text{ref},2}$ is generated from I_1 by a switchable NMOS current bank.

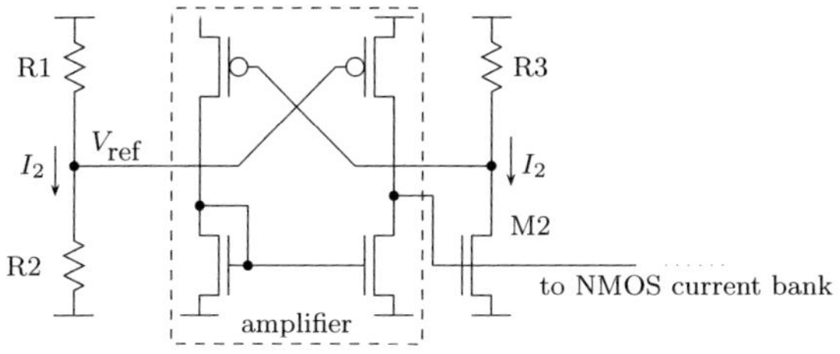

Figure 3.26: Current reference $I_{\text{ref},2,\text{pvk}}$, power-down switches not shown, [HHH+12]

3.2.4.4 Parameter Extraction

For a given circuit realization the reference current weighting factors $\mathbf{k} = (k_0, k_1, k_2)^T$ must be determined. This can be done either by circuit simulation or lab characterization of samples of the manufactured chips. The (V_{DD}, θ) characteristics of the DCO for several reference currents $T_{\text{DCO}} = F_1(V_{\text{DD}}, \theta, I_{\text{ref}})$ are determined. From this, the inverse characteristics $I_{\text{ref}} = |G(\theta, V_{\text{DD}})|_{T_{\text{DCO}}=T_0}$ are determined numerically by inverse interpolation. An example result is shown in Fig. 3.23. A two dimensional plane is fitted to G by the least square method using MATLAB which results in the linear approximation of the targeted bias current characteristics in Eq. 3.18. Thereby the vector $\mathbf{a}$ is determined. This linear characteristics must be reproduced by the bias circuit $I_{\text{ref}}(V_{\text{DD}}, \theta) = I_{\text{ref,comp}}(V_{\text{DD}}, \theta)$. Therefore, the three individual bias components are characterized for their linear (V_{DD}, θ) characteristics by circuit simulations or measurements, determining the matrix $\mathbf{A}$. Combining Eq. 3.21 with Eq. 3.18 and solving for $\mathbf{k}$ leads to

$$\mathbf{k} = \mathbf{a} \cdot \mathbf{A}^{-1}. \tag{3.29}$$

By determination of $\mathbf{k}$, the value for bias configuration signals cizero, ciptk and cipvk can be selected.

3.2.4.5 Implementation Results

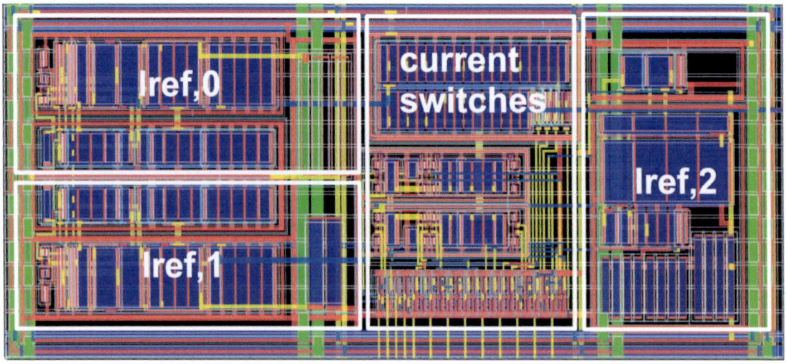

Figure 3.27: Layout of the DCO bias generator in 65nm CMOS, $24\mu\text{m} \times 54\mu\text{m}$

The bias current generator has been implemented in 65nm CMOS technology. Its layout is shown in Fig. 3.27. The robustness of the circuit with respect to process variations is evaluated using Monte-Carlo simulations. Tab. 3.2 shows the

Monte Carlo simulation results including global (process) and local (mismatch) variations (696 samples). The relative variability of the main current source parameters c_0, b_1 and a_2 are suitably low. Calibration of the bias circuit has been performed based on simulation data in the typical process corner (TT) using the method presented in Sec. 3.2.4.4. The resulting optimum settings for the bias current banks are $k_1 = 1.2$ and $k_2 = 0.9$ (normalized with respect to $k_0 = 1.0$). In this circuit implementation these settings are adjustable with 4-bit accuracy only. Thus there is a remaining weighting factor error which can be expressed by $\mathbf{k}/\mathbf{k}_{\text{ideal}} = (0.9848, -0.9045, -1.0395)$, i.e. the calibration of the linearized current bias is accurate within $\approx \pm 4\%$.

Table 3.2: Current source Monte-Carlo simulation results

	mean	**std**	**std/mean**
a_0 [μA/V]	-2.918	0.6985	-0.239
b_0 [μA/K]	0.05654	0.01129	0.200
$\mathbf{c_0}$ [μA]	100.49	13.00	0.131
a_1 [μA/V]	-0.5084	0.1407	-0.277
$\mathbf{b_1}$ [μA/K]	0.1149	0.01066	0.093
c_1 [μA]	41.97	4.069	0.097
$\mathbf{a_2}$ [μA/V]	16.33	1.116	0.068
b_2 [μA/K]	0.003123	0.002018	0.646
c_2 [μA]	2.97	0.207	0.070

Fig. 3.28 and Fig. 3.29 show the simulated and measured DCO periods respectively at fixed tuning values with and without bias current compensation. Tab.3.3 summarizes the results for period compensation of a DCO with fixed c_{coarse} and c_{fine} for $T_{\text{DCO}} \approx 0.5$ns.

Table 3.3: DCO period compensation results

(1.08V; 1.32V) (15°C; 85°C)	$\overline{T}_{\text{DCO}}$ [ns]	ΔT_{DCO} [ns]	$\Delta T_{\text{DCO}}/\overline{T}_{\text{DCO}}$
sim. uncomp.	0.503	0.196	39.0 %
meas. uncomp.	0.513	0.156	30.4 %
sim. comp.	0.499	0.036	7.4 %
meas. comp.	0.497	0.039	7.8 %

Fig. 3.30(a) and Fig. 3.30(c) show the simulated fine tuning curves for lock-in at $(V_{\text{DD}}, \theta)_{\text{best}}$ and $(V_{\text{DD}}, \theta)_{\text{worst}}$ respectively according to the illustration in Fig. 3.21. In order to evaluate the required fine-tuning range for application of this DCO within an ADPLL, the following measurement procedure is performed to cover all worst-case and best-case operation scenarios within the (V_{DD}, θ) range $V_{\text{DD,min}} \leq V_{\text{DD}} \leq V_{\text{DD,max}}$ and $\theta_{\text{min}} \leq \theta \leq \theta_{\text{max}}$ ranges as explained in Sec. 3.2.3.

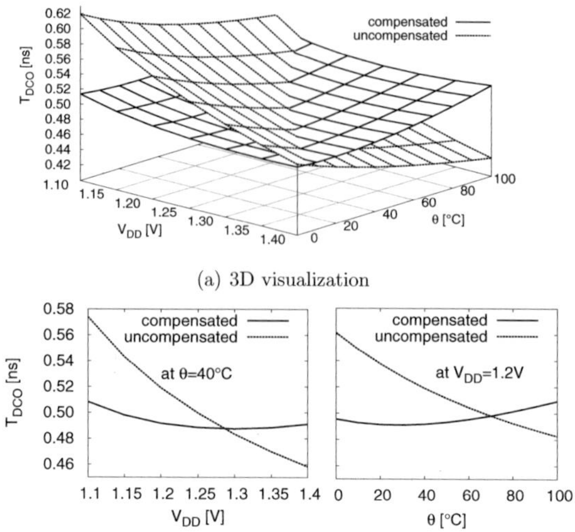

(a) 3D visualization

(b) 2D visualization

Figure 3.28: DCO period simulation results with (k_1 = 1.2 and k_2 = 0.9) and without ($k_1 = 0$ and $k_2 = 0$) compensated biasing, [HHH+12]

1. Initially lock the ADPLL at one ($V_{\mathrm{DD,min,max}}, \theta_{\mathrm{min,max}}$) corner
2. Apply resulting c_{coarse} to the DCO in open-loop mode
3. Measure $T_{\mathrm{DCO}}(c_{\mathrm{fine}})$ for three remaining ($V_{\mathrm{DD,min,max}}, \theta_{\mathrm{min,max}}$) corners
4. Repeat 1) to 3) for lock-in at all 4 ($V_{\mathrm{DD,min,max}}, \theta_{\mathrm{min,max}}$) corners

The resulting 16 tuning curves are plotted in Fig. 3.30(b) and Fig. 3.30(d) without and with bias compensation respectively. Measurement and simulation results are in good agreement. Without compensation for supply voltage and temperature variations it is not possible to maintain fine-lock for all possible changes of V_{DD} and θ during MPSoC operation. The bias compensation circuit allows to keep fine-lock during circuit operation with a maximum required c_{fine} range from 9 to 56 (simulation) and 14 to 55 (measurement), thereby enabling safe MPSoC operation under all specified V_{DD} and θ conditions.

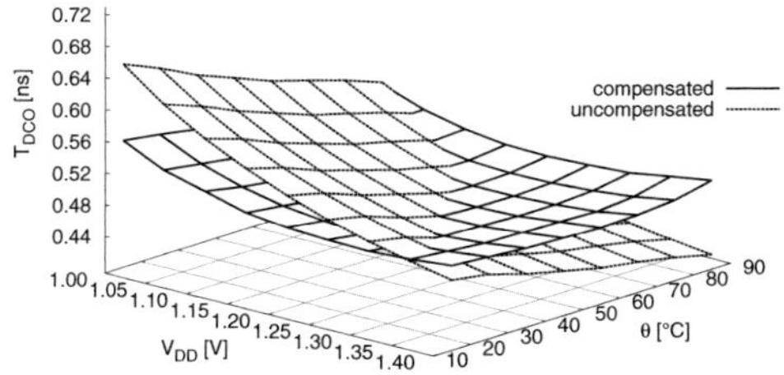

(a) 3D visualization

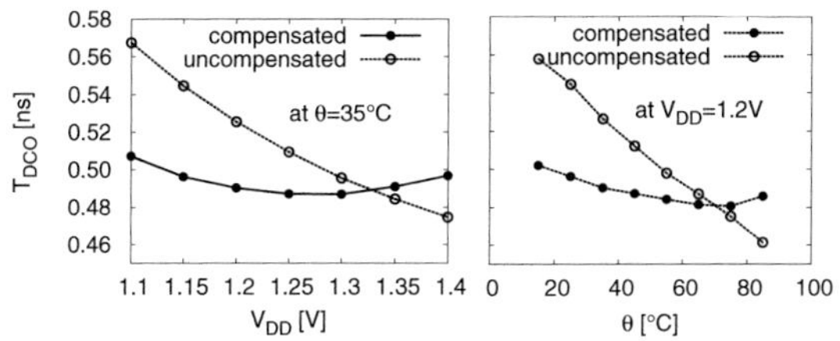

(b) 2D visualization

Figure 3.29: DCO period measurement results with ($k_1 = 1.2$ and $k_2 = 1.1$) and without ($k_1 = 0$ and $k_2 = 0$) compensated biasing, [IIIIII+12]

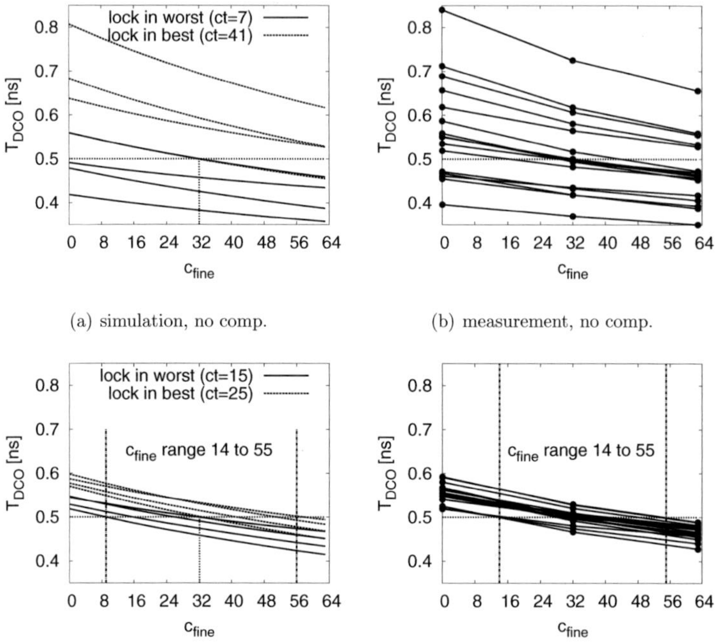

(a) simulation, no comp.

(b) measurement, no comp.

(c) simulation, comp.

(d) measurement, comp.

Figure 3.30: Fine tune lock-range results at $1.08\text{V} \leq V_{\text{DD}} \leq 1.32\text{V}$ and $0°C \leq \theta \leq 85°C$ ($15°C \leq \theta \leq 85°C$ for measurement)

3.3 A Multi-phase DCO in 28nm CMOS Technology

For the same target application, a multi-phase DCO with similar functionality and frequency range as presented in Sec. 3.2 is implemented in a leading-edge 28nm CMOS technology. Due to the fact that short channel effects with the reduction of the MOS transistor output resistance as well as device variability increase significantly in smaller CMOS technologies, the implementation of a supply voltage and temperature compensation scheme as presented in Sec. 3.2 is challenging. The size of the transistors would have to be increased significantly with respect to the minimum feature size of the technology node. This counteracts the area savings which come from technology scaling, and are mandatory to realize a DCO circuit which is ultra-compact for per-core instantiation within MPSoCs. Therefore a different tuning technique is used for this 28nm DCO implementation. A digitally controllable series resistance in the supply path is used for tuning as presented in Sec. 3.1.1.5. The tuning resistor is built up using PMOS devices connected in parallel [KSK+09]. This purely digital tuning scheme without analog voltage or current processing elements benefits well from technology scaling in terms of area, does not rely on good analog MOS device properties and allows for easy design implementation.

3.3.1 Circuit Overview

The oscillator core architecture as shown in Fig. 3.31(a) is similar to the one of the 65nm DCO realization presented in Sec. 3.2. It consists of tristate inverter cells (Fig. 3.31(b)), which can be disabled during power down. The cross-coupled inverters in each stage are not completely disabled, but the pull-up (p on) and pull-down paths (n on) remain on during power down, realizing a reset scheme, where the differential ring oscillator nodes keep their differential state all times. This ensures safe and defined startup conditions as analyzed in Sec. 3.2.
The tuning circuit is shown in Fig. 3.32. The employed tuning scheme with a digitally controlled resistor in the supply voltage net has been analyzed in Sec. 3.1.1.5. The supply tuning resistance is realized by parallel connection of PMOS devices, which are encapsulated into tune switch cells. Each tune switch cell contains a HVT PMOS switch with $4 \cdot L_{\text{min}}$ gate length, to achieve a higher on-resistance for finer tuning step size as motivated in Sec. 3.1.1.5. It is enabled by a standard CMOS inverter structure, which additionally serves the purpose of decoupling the digital supply voltage domain (where the ADPLL controller is located) from the analog DCO supply voltage domain, to reduce supply noise on the analog domain. The tune switch cell layout is compatible to the standard cell grid. The remaining layout

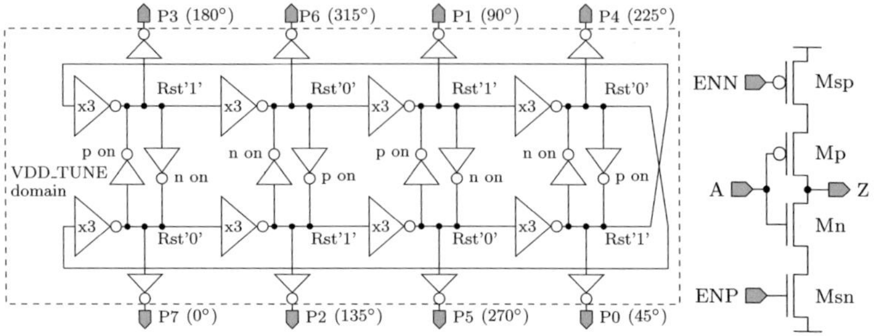

(a) ring oscillator (b) core cell

Figure 3.31: 28nm DCO core schematic

space for NMOS due to the PMOS switch with increased length is filled with an capacitor device, which reduces ripple on the DCO tuning voltage. Each single tune switch has the on conductance of $G_0 = 1/R_{\text{on,HVT,PMOS}}$. The tune switch cells are clustered for different components of the total tuning signal c of 10-bit width. The main group are 64 tune switch cells which are controlled by 64 thermometer coded signals cTHERM0 to cTHERM63, representing the upper 6 bits of c. The 4 LSBs of c are directly applied to 4 tune switch cell with binary increased on-resistances by series connection of the high threshold voltage (HVT) PMOS devices, realizing on conductances of $G_0/2$, $G_0/4$, $G_0/8$ and $G_0/16$, respectively. Another switch with $G_0/16$ is added for the delta sigma modulated LSB tune bit $c_{\text{DSM}} \in [0;1]$. In order to achieve a small tuning step size and to allow oscillation of the DCO for all possible values of c_{tune}, a set of 20 always-on switch cells is added. The number of the activated always-on cells is selectable by c_{ao}, which is not changed during ADPLL control of this DCO. The total tune conductance in the supply path reads

$$G_{\text{tune}} = c_{\text{ao}} \cdot G_0 + c \cdot \frac{G_0}{16} + c_{\text{DSM}} \cdot \frac{G_0}{16} \tag{3.30}$$

where c_{DSM} is an additional LSB which is used for fractional tuning using a delta-sigma modulator (DSM) as presented in Sec. 4.1.3.2.

3.3.2 Implementation Results

Fig. 3.33 shows the layout of the DCO in 28nm CMOS technology. All components are realized within a standard cell grid. The VDD supply rail of the oscillator core cells is connected to the VDD_TUNE net, whereas the supply rails of the tuning switch array and the output buffers are operating on VDD. Therefore adapter cells

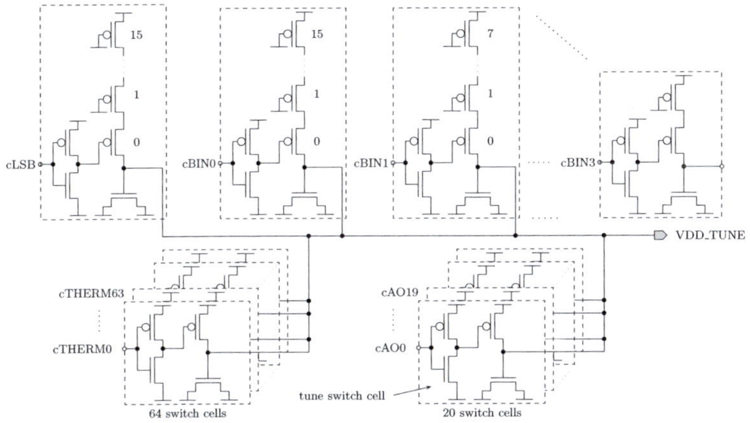

Figure 3.32: Tuning circuit schematic with resistance in the supply path

are inserted left and right of the ring oscillator core, which also realize required spacing of the n-wells on VDD and VDD_TUNE, respectively.

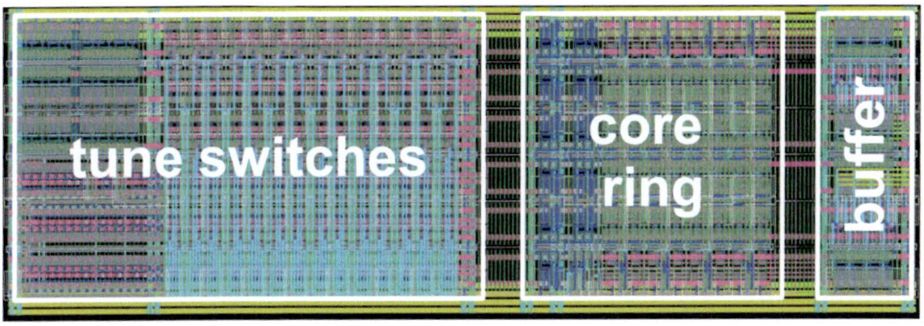

Figure 3.33: Layout of the DCO in 28nm CMOS technology, 28.0μm × 8.6μm

Fig. 3.34 shows the simulated tuning characteristics of the DCO within the specified temperature and supply voltage ranges ($-40° \leq \theta \leq 125°$, $0.9\text{V} \leq V_{\text{DD}} \leq 1.1\text{V}$) for three process corners. The target period of $T_0 = 0.5$ns can always be achieved within these ranges. The tuning gain $K_t = \text{d}T_{\text{DCO}}/\text{d}c_{\text{tune}}$ shows strong dependency on the PVT conditions. In the lower plots of Fig. 3.34 the target K_t at $T_{\text{DCO}} = T_0$ is printed. It shows variations from 0.3ps to 2.4ps considering all possible PVT variations within the specified ranges.

Fig. 3.35 shows the measured output waveform and the period jitter histogram of the DCO when running at ≈ 2GHz. The DCO signal is fed through an LVDS pad as described in Sec. A.3. Fig. 3.36 shows example measurement results of T_{DCO}

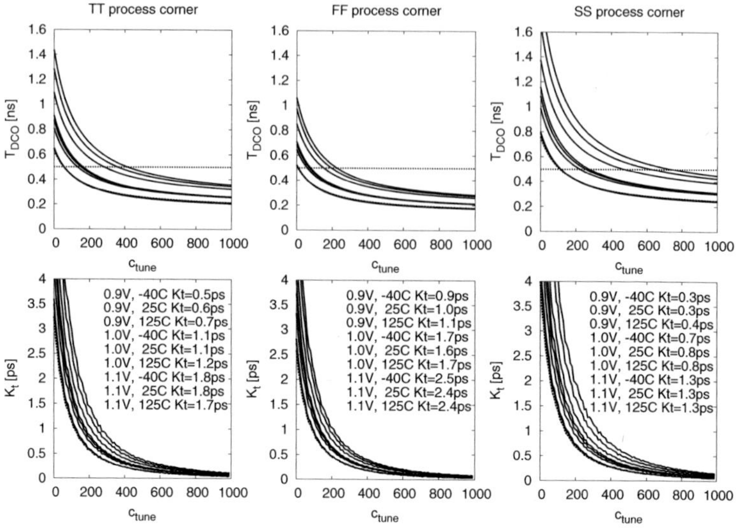

Figure 3.34: DCO tuning characteristics and tuning gain, $-40° \leq \theta \leq 125°$, $0.9\text{V} \leq V_{\text{DD}} \leq 1.1\text{V}$

and the supply current I_{DD} for different numbers of activated always-on switches. Tab. 3.4 summarizes the performance of the DCO in 28nm CMOS technology. The achievable tuning range does not include the configuration of the always on switches, which can be adjusted, if the target frequency of the DCO has to be in another frequency range than 2GHz.

Table 3.4: 28nm DCO performances, post-layout simulation results, corners

	min	**typical**	**max**
T_{DCO} [ps] achievable over all PVT conditions	450	500	515
T_{DCO} [ps] achievable over all PT conditions at 1V	312	500	645
K_{t}=T_{step} [ps] at 2GHz	0.3	1.6	2.4
period jitter rms (simulated, w/o supply noise) [ps]	0.62	0.95	0.95
period jitter rms (measured, see. App. A.3) [ps]		3	
duty cycle [%]	40	48	55
power consumption [mW] at 2GHz	0.27	0.36	0.59

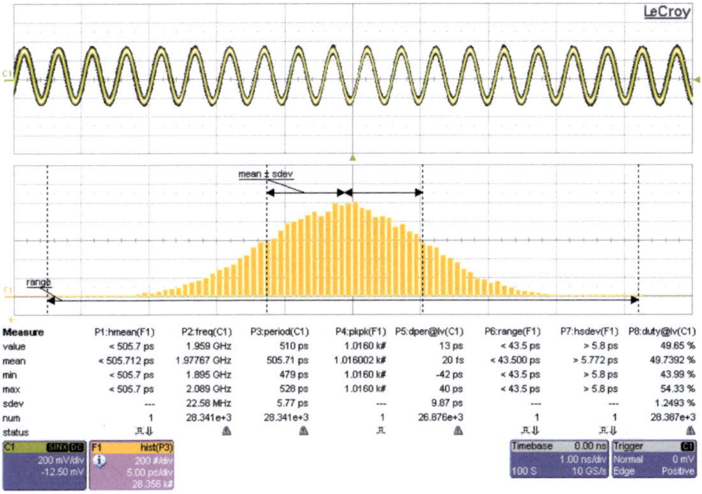

Figure 3.35: Measured 28nm DCO output waveform over LVDS pad

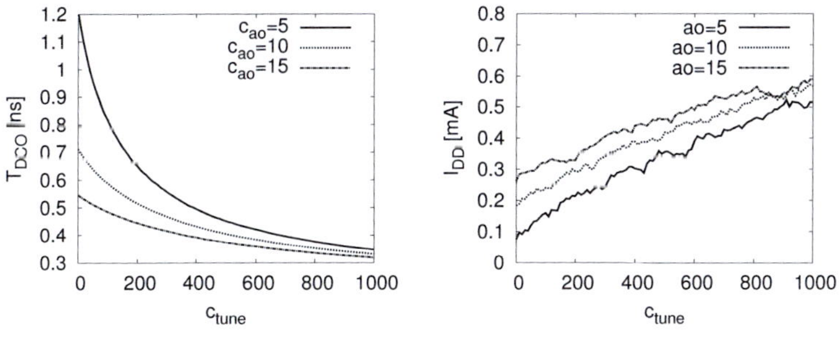

(a) tune

(b) current consumption

Figure 3.36: 28nm DCO measurement results, $V_{DD} = 1.0\text{V}$, room temperature

3.4 Differential Clock Buffers

The DCO topologies that have been presented in Sec. 3.2 and Sec. 3.3 provide multi-phase and differential output clock signals, which can be used for open-loop clock generation circuits as shown in Sec. 5.1 or for double-data-rate transmission circuits as presented in Sec. 2.4. The key signal properties are a well defined phase shift between the clock phases and a good duty cycle near 50%. These properties have to be maintained when distributing these clock signals from the generator to its target circuit or when crossing voltage domains. This can be achieved by the application of differential clock buffers based on cross-coupled converters, described in the following subsections. The phase allignment properties of this circuit topology are analyzed theoretically.

3.4.1 Circuit overview

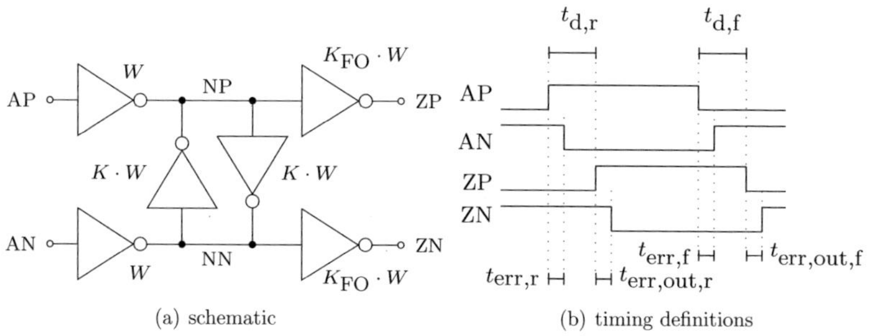

Figure 3.37: Differential buffer with cross-coupled inverters

Fig. 3.37(a) shows the schematic of the differential clock buffer with cross-coupled inverters. It consists of symmetric input inverters [2] with a nominal device width W. Cross-coupled inverters with a nominal drive strength of $K \cdot W$ provide positive feedback between the differential internal circuit nodes NP and NN. The output nodes are driven by inverters with higher drive strength of $K_{\mathrm{FO}} \cdot W$.
Fig. 3.37(b) illustrates the timings of the differential clock buffer. The input signals at the nodes AP and AN may exhibit a timing mismatch between their rising and falling edges of $t_{\mathrm{err,r}}$ and $t_{\mathrm{err,f}}$, respectively. For an ideally differential input signal these timing errors are zero. The delay through the differential clock buffer is $t_{\mathrm{d,r}}$

[2] As these circuits are used for driving clocks, the ratio between the PMOS width W_{P} and NMOS width W_{N} is chosen such that the rising edge delays and falling edge delays through the single inverter are as equal as possible

and $t_{d,f}$ for the positive input node. In the following analyses the timing errors and delays for only one signal edge are considered. Results can be directly referred to the other edge due to the symmetry of this circuit structure.

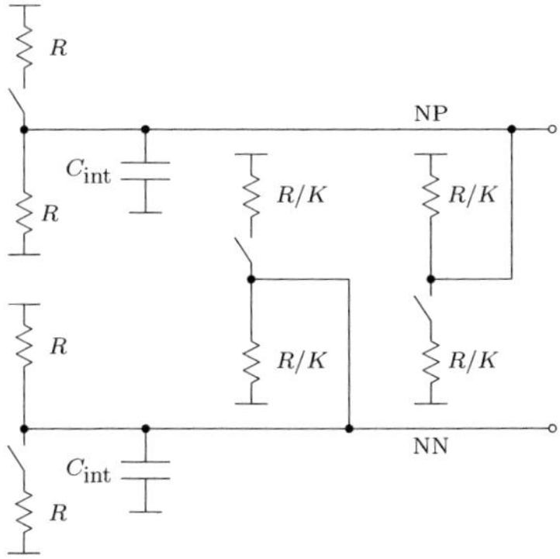

Figure 3.38: Equivalent RC schematic of the differential clock buffer, directly after rising edge at AP with AN=0

For principle analysis of this circuit topology, the inverters are modeled as switched resistances [Uye01]. For simplicity we consider the clock inverters to be balanced with $R_{on,PMOS} = R_{on,NMOS} = R$. It is assumed that the inverters switch if the input voltages reaches $V_{DD}/2$. Thereby, the internal timings can be calculated by solving the linear differential equations of these RC-charging processes. The general time domain solution of an RC charging process on a single capacitance reads

$$V(t) = (V(0) - V(\infty)) \cdot e^{-\frac{t}{\tau}} + V(\infty) \tag{3.31}$$

where $V(0)$ and $V(\infty)$ are the capacitor voltages at the beginning of the charging process and after a long time $t \to \infty$. As example, Fig. 3.38 shows the equivalent schematic of the differential clock buffer, directly after a rising edge occurred at AP. First the nominal delay time through the differential clock buffer is calculated, assuming that the signals are in ideal phase, i.e. $t_{err} = 0$. Only the half circuit is considered here. When the internal nodes, starting from $V(0) = 0$ do not yet have reached the switching threshold, the cross-coupled inverters drive on the opposite node. The target voltage of the linear RC charging process is $V(\infty) = V_{DD}/(1+K)$. The RC time constant of the internal nodes NN and NP reads

$$\tau = \frac{1}{1+K} \cdot R \cdot C_{int} \tag{3.32}$$

where C_{int} is the internal capacitance including the input capacitance of the output inverter. It is

$$C_{\text{int}} = C' \cdot (K + K_{\text{FO}}) \cdot W + C_{\text{par}}. \tag{3.33}$$

The nominal delay, which is the sum of the charging time of the internal node to $V_{\text{DD}}/2$ from Eq. 3.31 and the additional delay of the output inverter, results in

$$t_{\text{d}} = \tau \cdot \ln\left(\frac{2}{1-K}\right) + \frac{1}{K_{\text{FO}}} \cdot R \cdot C_{\text{L}} \cdot \ln 2 \tag{3.34}$$

where C_{L} is the output load capacitance. This result indicates, that the nominal delay increases if the cross-coupling factor K increases.

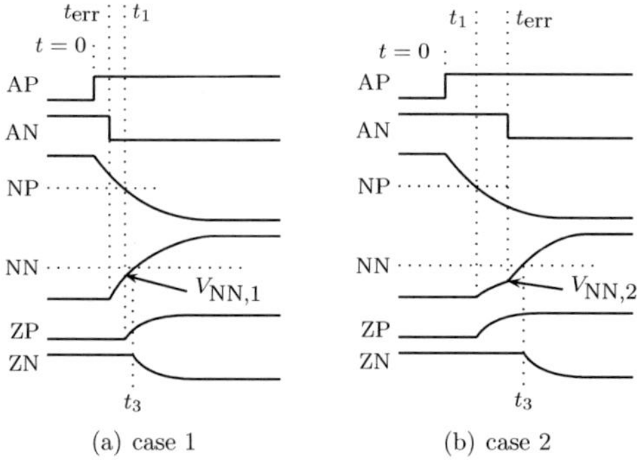

Figure 3.39: Differential clock buffer timings with input timing error

In order to calculate the propagation of a timing error t_{err} through this differential clock buffer structure, Fig. 3.39 gives a more detailed view on the timings if an input timing error is present. Triggered by the rising edge at input AP, the internal node NP is driven against the cross-coupling and reaches its threshold level after time t_1, which triggers the switching of the first output node ZP. This causes the node NN to be pre-charged by the cross-coupled inverters. The falling edge at input AN occurs after t_{err}. Here two cases are distinguished:

In *case 1*, the falling edge at AN occurs before the first cross-coupled inverter switches, i.e. $t_{\text{err}} \leq t_1$. At t_1 the node NN has been pre-charged to $V_{\text{NN},1}$, from which the final charging process to full swing CMOS level starts.

In *case 2*, the falling edge at AN occurs after the first cross-coupled inverter switches, i.e. $t_{\text{err}} > t_1$. At t_{err} the node NN has been pre-charged to $V_{\text{NN},2}$, from which the final charging process to full swing CMOS level starts.

The node NN reaches the threshold at time t_3, which triggers the switching of the second output node ZN. Thus the remaining timing error at the output can be expressed by

$$t_{\text{err,out}} = t_3 - t_1. \tag{3.35}$$

By the pre-charging of the internal node NN, its transition to the full swing CMOS level is accelerated, which leads to an reduced output timing error $t_{\text{err,out}} < t_{\text{err}}$. Fig. 3.40 shows Spectre simulation results of a cross-coupled inverter buffer in 65nm CMOS (complete transistor models) for the two cases as mentioned above.

The remaining output timing error $t_{\text{err,out}}$ can be calculated by solving the RC differential equations of the charging processes of nodes NN and NP, respectively. All calculations are performed for $t_{\text{err}} > 0$, but are also valid for $t_{\text{err}} < 0$ due to the symmetry of the differential clock buffer topology. The time after node NP reaches its threshold is (for case 1 and case 2)

$$t_1 = \tau \cdot \ln\left(\frac{2}{1-K}\right) \tag{3.36}$$

with τ from Eq. 3.32. In case 1 it is the asymptotic voltage of the charging process to $V_{\text{NN},1}$

$$V_{\text{NN},1,\infty} = \frac{1}{1+K} \cdot V_{\text{DD}} \tag{3.37}$$

because the input inverter is driving against the cross coupling inverter which holds the previous signal level. In case 2 it is

$$V_{\text{NN},2,\infty} = \frac{K}{1+K} \cdot V_{\text{DD}} \tag{3.38}$$

because the cross couple inverter is driving against input inverter. Therefore with Eq. 3.31 the values of the pre-charge voltage of NN at t_1 (case 1) and t_{err} (case 2) read

$$\frac{V_{\text{NN},1}}{V_{\text{DD}}} = \frac{1}{1+K} \cdot \left(1 - e^{-\frac{t_1 - t_{\text{err}}}{\tau}}\right) \tag{3.39}$$

$$\frac{V_{\text{NN},2}}{V_{\text{DD}}} = \frac{K}{1+K} \cdot \left(1 - e^{-\frac{t_{\text{err}} - t_1}{\tau}}\right). \tag{3.40}$$

From this the time t_3, when NN reaches $V_{DD}/2$ can be calculated. It is for case 1

$$\frac{1}{2} = \left(1 - \frac{V_{NN,1}}{V_{DD}}\right) \cdot e^{-\frac{t_{3,1}-t_1}{\tau}} \tag{3.41}$$

and for case 2

$$\frac{1}{2} = \left(1 - \frac{V_{NN,2}}{V_{DD}}\right) \cdot e^{-\frac{t_{3,2}-t_{err}}{\tau}}. \tag{3.42}$$

From this, the remaining output timing error can be calculated using Eq. 3.35 and Eq. 3.36. It results in

$$t_{err,out}(K) = \begin{cases} \tau \cdot \ln\left(2 - 2 \cdot \frac{V_{NN,1}}{V_{DD}}\right) & ;\ t_{err} \leq t_1 \\ t_{err} + \tau \cdot \ln\left(2 - 2 \cdot \frac{V_{NN,2}}{V_{DD}}\right) - \tau \cdot \ln\left(\frac{2}{1-K}\right) & ;\ t_{err} > t_1. \end{cases} \tag{3.43}$$

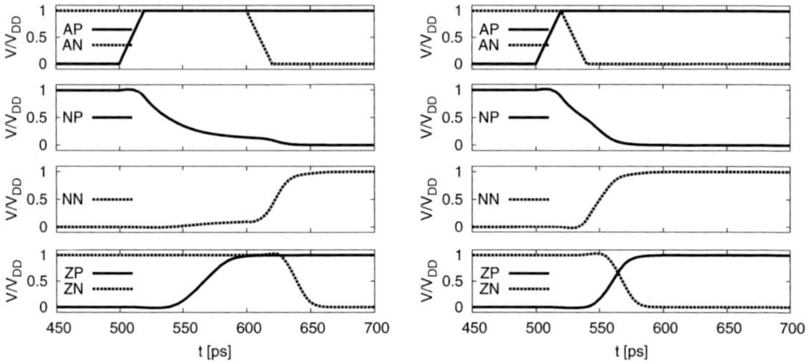

(a) Case 1 (b) Case 2

Figure 3.40: Differential clock buffer simulation results waveforms, 65nm CMOS

Fig. 3.41(a) visualizes the result from Eq. 3.43 together with Spectre simulation results from a differential clock buffer circuit in 65nm CMOS technology. For $k < 1.0$ the results are in good agreement. Fig. 3.41(b) shows the simulated energy per toggle of the differential clock buffer from Spectre simulations. It emphasizes the trade-off between the ability of timing error reduction with larger values of k and the increased energy consumption due to the larger cross-coupling devices.

The analysis results illustrate that the differential clock buffer topology with cross-coupled inverters, can reduce phase errors in differential clock signals effectively. The results in Eq. 3.34 and Eq. 3.43 indicate that a strong cross coupling with $k \to 1$

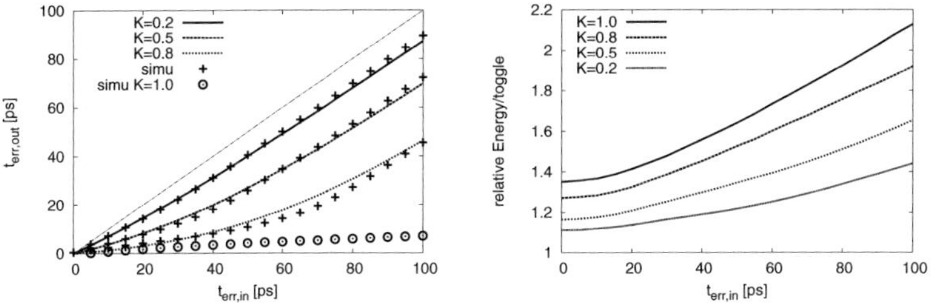

(a) timing error transfer function (Eq. 3.43), analytical solution and Spectre simulation results, $C_{\text{par}} = 5\text{fF}$, $C' = 5\text{fF}/\mu\text{m}$, $R = 2.5\text{k}\Omega$

(b) energy per toggle, Spectre simulation results, normalized with respect to energy for $K = 0$

Figure 3.41: Differential clock buffer Spectre simulation results, 65nm CMOS, $K_{\text{FO}} = 2$, $W = 1\mu\text{m}$, $C_{\text{L}} = 10\text{fF}$

is not feasible, although it provides excellent timing error reduction. However, for practical applications values of $k = 1$ can be applied because the drivestrenght of the cross-coupling inverters reduces when changing their input voltage by pre-charging of the internal nodes NP and NN, which is not considered in the simplified linear model presented above. Anyway, the risk of a latching effect, where the state of the cross-coupled inverters can not be switched by the input inverters increases for large k. Therefore these cross-coupled clock buffers must be verified carefully by Monte-Carlo simulations considering mismatch. In the circuit implementations in this work, a value of $k = 1$ is chosen.

3.4.2 Duty cycle adjustment of multi-phase clock signals

The 28nm DCO realization with resistive tuning in the supply voltage path as shown in Sec. 3.3, suffers from duty cycle distortion as illustrated in Fig. 3.42. The multiphase clocks cross the voltage domain from the inner $V_{\text{DD,tune}}$ to the nominal supply voltage level. Due to the fact that $V_{\text{DD,tune}} < V_{\text{DD}}$, the NMOS devices N1 are not switched completely at a rising edge at its input. Therefore the rising edge delay through this buffer structure is larger than the falling edge delay, leading to a reduced duty cycle. Because all clock phases are effected by the selective rise and fall delays in the same manner, this does not disturb the phase relation between the clock phases. Therefore no additional duty cycle compensation is required, if the clocks are only used with respect to their rising or falling edge exclusively. This is the case for the open-loop clock generator based on phase switching and frequency division as presented in Sec. 5.1. If both clock phases are to be used, e.g. for clocking double data rate (DDR) data transmission circuits (see Sec. 2.4), duty cycle adjustment is mandatory.

Ideally, the falling edge of one clock phase corresponds to the rising edge of the 180 degree shifted clock. The common duty cycle distortion of all 8 clock phases can be interpreted as timing error t_{err} between the clocks being 180° out of phase, according to the definition in Sec. 3.4.1. If both are combined by the differential clock buffer with cross-coupled inverters, the duty cycle can be improved while maintaining the phase difference.

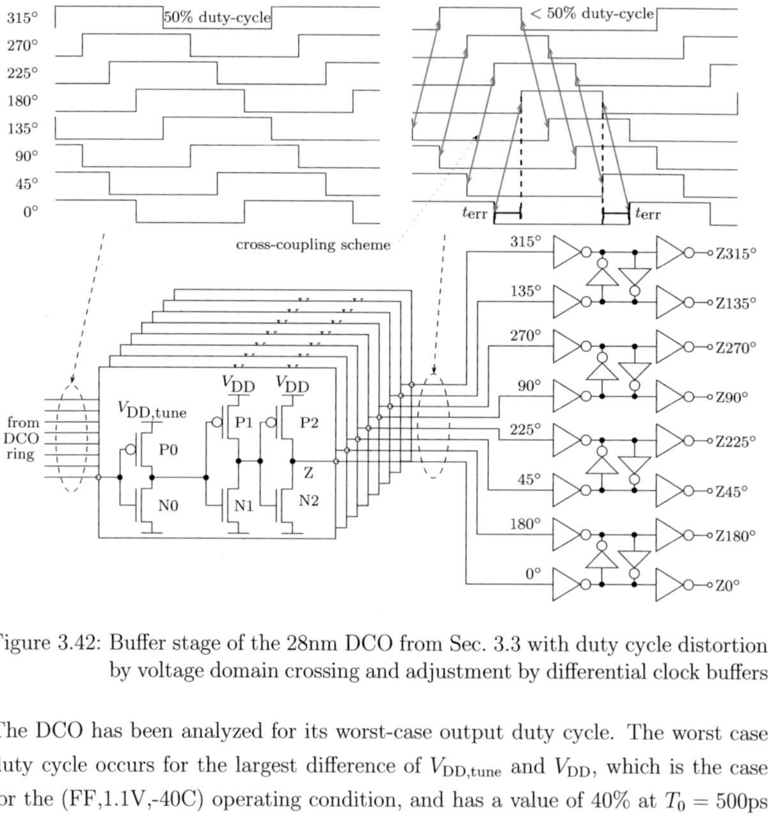

Figure 3.42: Buffer stage of the 28nm DCO from Sec. 3.3 with duty cycle distortion by voltage domain crossing and adjustment by differential clock buffers

The DCO has been analyzed for its worst-case output duty cycle. The worst case duty cycle occurs for the largest difference of $V_{\text{DD,tune}}$ and V_{DD}, which is the case for the (FF,1.1V,-40C) operating condition, and has a value of 40% at $T_0 = 500$ps nominal period. Differential clock buffers with cross-coupled inverters (with parameters $k = 1$, $k_{\text{FO}=2}$) have been applied to the multiphase clocks to improve this duty cycle. Fig. 3.43 shows the simulated duty cycle correction from post-layout views for typical, best and worst operating conditions. In the worst-case duty cycle scenario (best case timing conditions) the minimum output duty cycle can be improved from 40% to 46%.

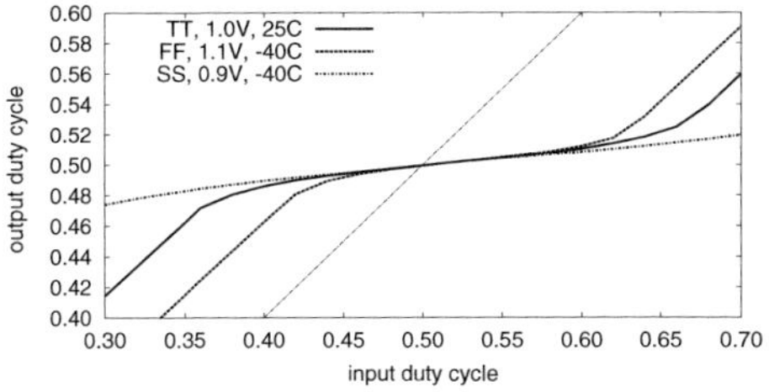

Figure 3.43: Duty cycle correction simulation results, 28nm clock buffers, post-layout simulation, three process corners (TT,FF,SS)

Robustness of the clock buffer circuit with respect to process variations is analyzed by Monte-Carlo simulations including global and local variations for the worst-case input duty cycle of 40%. Fig. 3.44(a) show the results histogram of one of the 8 clock phases as example. The phase relation of the 8 clock phases is not disturbed by the differential clock buffers. At $T_0 = 500$ps the nominal skew between adjacent phases is 62.5ps. Monte-Carlo simulations show for each skew an average value of nearly 62.5ps and a maximum standard deviation of 3.1ps. As example, Fig. 3.44(b) shows the skew histogram from 1000 Monte-Carlo runs between phases 0° and 45°. This proves the robustness of the proposed differential clock buffer even with a sizing of $k = 1$ for small CMOS technologies with significant process variability.

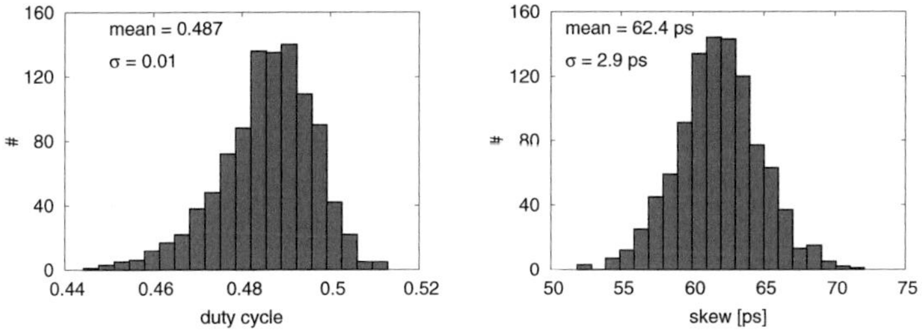

(a) duty cycle of phase 0°

(b) skew between phase 0° and phase 45°

Figure 3.44: Statistical simulation results of the differential clock buffer output of the 8-phase DCO in 28nm CMOS, Monte-Carlo result from 1000 runs, global and local variations, input duty cycle 40%, $V_{DD} = 1.0V$

3.5 Summary

Ring oscillator DCOs are the key components of ADPLL based clock generators. Various types of digital tuning mechanisms have been reviewed. They are often used in combination to achieve both wide tuning range for PVT compensation and a small tuning step size for low jitter operation. Two different digitally controlled multi-phase oscillators have been developed in 65nm and 28nm CMOS technology for the application in this work. Both have a target frequency of 2GHz.

The 65nm realization employs a current-starved inverter topology with a centralized DAC tuning. Separated coarse and fine tuning functionality is available. A novel current bias generator is presented which compensates the supply voltage and temperature dependency of the ring oscillator, which enables operation within the specified operating condition ranges without employing the coarse tune scheme after lock-in. This prevents additional circuit effort to calibrate the monotony of the DCO characteristics with respect to the coarse and fine tuning word, as it has been presented previously.

The 28nm realization uses a programmable resistor in the supply net to tune the core ring. For fine resolution of 10-bit hundreds of PMOS switches are used here. Since this circuit does not require any analog bias components it benefits well from the aggressive scaling of purely digital cells in this 28nm CMOS node.

Both DCOs fulfill the requirements for the targeted MPSoC clocking applications in terms of frequency, power consumption, period jitter and robustness with respect to PVT variations.

As required for distribution of differential and multi-phase clock signals in the targeted application, differential clock buffers with cross-coupled inverters have been developed and analyzed theoretically. They allow efficient compensation of phase mismatch caused by process variations.

4 All-digital Phase-locked Loops

PLLs can be used to control oscillators for generation of defined frequencies from a reference clock signal and to suppress their jitter accumulation. This chapter introduces the PLL circuit concept with special focus on its minimalistic all-digital realization (bang-bang ADPLL (BBADPLL)). A numerical system model is developed to analyze the BBADPLL system performances with respect to design parameters in both the digital controller components and the DCOs as presented in the previous chapter. From this an ADPLL controller with fast clocking is proposed to reduce the jitter accumulation. ADPLLs are realized in both 65nm and 28nm CMOS technologies. For ultra-fast lock in a novel phase synchronization technique is developed, analyzed theoretically and implemented in 28nm CMOS technology.

4.1 Circuit Architecture

4.1.1 Overview

Based on their circuit realization in terms of digital and analog components, two mayor types of PLLs are commonly used for MPSoC clocking applications. Charge pump PLLs, as shown in Fig. 4.1(a), employ a voltage controlled oscillator (VCO). The oscillator clock is divided by the loop divider and is compared to the reference clock by the PFD which produces control signal outputs up and down. A charge pump (CP) adds or removes charge of the analog loop filter where the amount of charge depends on the phase difference at the PFD input. The phase detection characteristics are shown in Fig. 4.2(a). Thereby the VCO frequency and phase is locked to the reference clock. Due to the continuous VCO tuning and passive filter structures charge pump PLLs can provide low jitter output clocks at reasonable power consumption [SSS05, Fah05, YYG08, HL09, CL10, CL09], even in modern nanometer CMOS technologies [PKGN12]. This also enables application for wireless communication frequency generation, as for example fractional-N frequency synthesis [Höp08]. However, the design realization of low noise charge pump PLLs requires the consideration and optimization of noise contribution of various analog circuit components [GKGN09]. Additionally these analog realizations suffer from the reduced voltage headroom in modern CMOS technologies with lower supply voltages [HSTW10]. This complicates design implementation and increases the effort of transferring a PLL to a new technology node. Another drawback is the need for off-chip passive loop filter components or a large chip area (defined by the available capacitance per-area of the technology), when integrating the passive loop filter completely on-chip [SEH+12].

In contrast to that, ADPLLs as shown in Fig. 4.1(b) use a DCO. A time-to-digital converter (TDC) digitizes the phase difference between the reference clock and the divided oscillator clock as shown in Fig. 4.2(b). In case a one-bit TDC is used, which simply outputs the sign of the input phase difference as shown in Fig. 4.2(c), the circuit is called bang-bang PFD (BBPFD). The loop filter is realized as digital circuit, producing the digital tuning signal for the DCO.

ADPLLs show a lot of benefits compared to their analog counterparts. The chip area can be reduced significantly due to the digital realization of the loop filter, especially in nanometer CMOS technologies [HEH+13, CSM10, TRF08]. The digital nature of tuning and control signals provides a lot of flexibility for configuration during operation (e.g. adaptive loop filter bandwidth) or for test and measurement purposes. Implementation of the main ADPLL circuit parts can be performed efficiently using standard digital implementation flows. However, the design and implementation of DCOs is usually more complex compared to VCOs. Both show the same relation between power and noise [GKGN09], but in case of the DCO also the digital tuning circuitry with trade-offs in terms of tuning range and tuning step size must be considered. Details on this have been given in Sec. 3.1.

Generally, ADPLLs are well suited for integration in nanometer CMOS technologies, where digital circuits shrink well in terms of chip area and power consumption, whereas the integration of analog components gets complicated due to imperfectness of semiconductor devices (e.g. short channel effects).

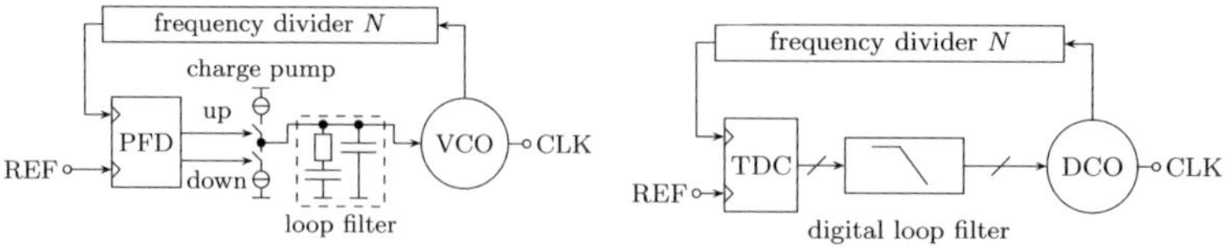

(a) charge pump PLL with analog filter (b) All-digital PLL

Figure 4.1: Types of PLL frequency synthesizers

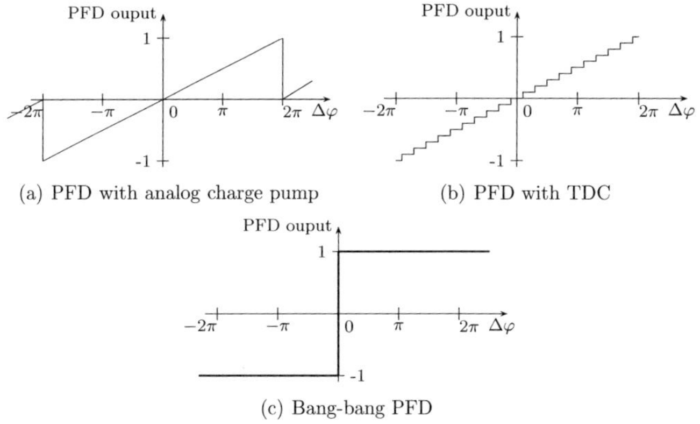

(a) PFD with analog charge pump

(b) PFD with TDC

(c) Bang-bang PFD

Figure 4.2: Types of phase-frequency detectors (PFDs)

4.1.2 Bang-bang ADPLL

Bang-bang ADPLLs are ideally suited for MPSoC clocking applications because no complex multi-bit TDC phase detectors or counter based frequency detectors are required, which eases design implementation and reduces power consumption. The BBPFD is a simple digital asynchronous circuit, which performs a binary comparison of the reference clock phase and the divided DCO signal [TRF08]. Details of the BBPFD implementation in this work are given in Sec. 4.2.1.4.

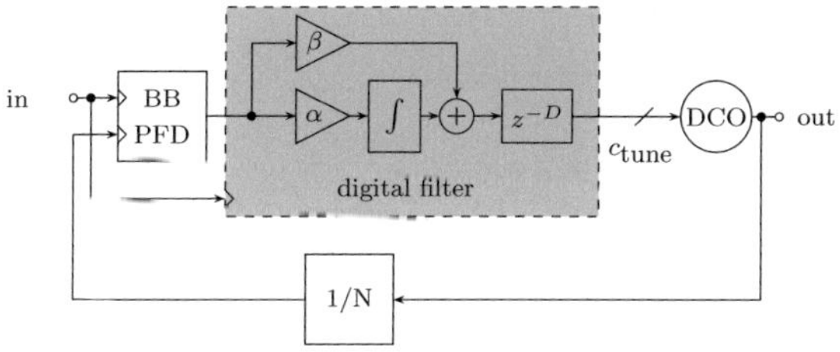

Figure 4.3: Bang-bang ADPLL schematic

Fig. 4.3 shows the basic schematic structure of a BBADPLL. The one-bit PFD output is filtered by a digital filter consisting of a proportional path with gain β and an integral path with gain α. The digital filter exhibits a delay of D clock cycles for computation of the DCO tuning word. The DCO has a linearized tuning gain of K_t with

$$K_{\mathrm{t}} = \frac{\mathrm{d}T_{\mathrm{DCO}}}{\mathrm{d}c}. \tag{4.1}$$

The basic BBADPLL is a nonlinear system with two state variables. The first one is the value of the integrator register ϕ (representing the averaged tuning word and thereby the frequency of the DCO) and the second one is the timing difference of the reference signal edge and the divider signal edge at the PFD input (representing the phase accumulation of the DCO). Therefore the BBADPLL dynamics can be visualized in a two-dimensional state space plot as shown in Fig. 4.10(b).

The BBADPLL dynamics have been analyzed mathematically in [DD05]. To ensure that the bang-bang loop can lock-into a bounded orbit the relation

$$\frac{\alpha}{\beta} < \frac{2}{2D+1} \tag{4.2}$$

must be fulfilled. In this case the deterministic jitter due to the non-linear loop behavior can be calculated [DD05].

The BBADPLL serves two purposes. First, the DCO output clock signal must have a constant average period of T_{ref}/N. When starting the ADPLL the time that is required to achieve phase and frequency lock is defined as t_{lock}. Second, the jitter which is accumulated within the noisy DCO oscillation loop has to be attenuated by locking to the reference clock signal with low jitter (e.g. from a crystal oscillator) [MR09]. This is especially important when employing the ADPLL for DDR memory interfaces, which specify limits for long term accumulated jitter [JED07, JED09, JED10]. Details on different jitter metrics are summarized in App. A.1.

Fig. 4.4 illustrates the phase noise transfer behavior of the closed-loop ADPLL. This frequency domain analysis can be efficiently applied to study the noise shaping of BBADPLLs by employing linearized models of their components [DD08, ZTL+09, PK13]. The phase noise of the reference clock is low-pass filtered by the loop filter, whereas the DCO phase noise is high-pass filtered by the loop, i.e. low frequency components of DCO phase fluctuations are suppressed.

Translating this to time domain, as visualized in Fig. 4.5, describes the jitter transfer behavior of the loop. The free running DCO jitter accumulation is suppressed which results in limited accumulated jitter after a number of cycles n. Since the loop filter of the BBADPLL operates with the low frequency reference clock, the DCO tuning word c remains constant for several DCO cycles depending on the

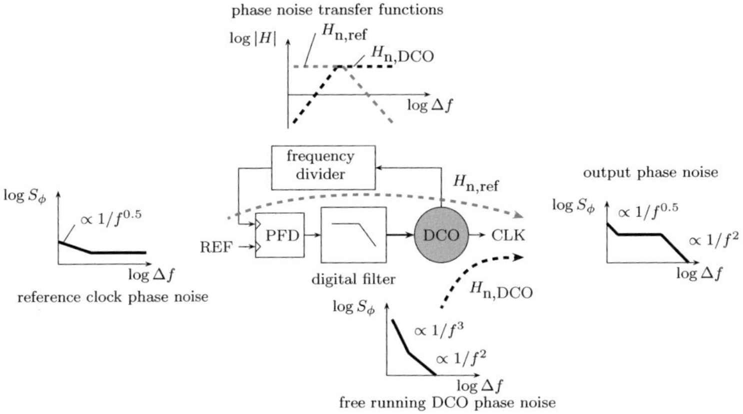

Figure 4.4: ADPLL noise transfer

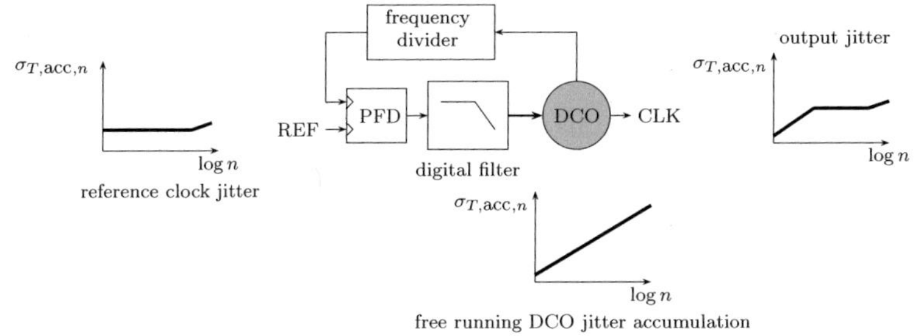

Figure 4.5: ADPLL jitter transfer

loop division ratio N. Therefore *period jitter* of the DCO can not be attenuated by the closed loop ADPLL. It can even be increased by discrete tuning, if the DCO tuning step size T_{step} is in the range of the period jitter standard deviation $\sigma_{T,\text{DCO}}$. In contrast, the long term phase error accumulation due to DCO noise is attenuated by the BBADPLL. This jitter transfer behavior is analyzed in detail in [DD08, ZTL+09, PK13]. Therefore the loop is modeled by an equivalent linear small signal model. The equivalent finite gain of the BBPFD in this case depends on the jitter at its input [DD06]. Analytical solutions of the integrated PLL jitter in terms of random jitter due to DCO noise and deterministic jitter due to the orbit of the nonlinear control loop are provided [ZTL+09]. These results indicate that there exist optimum settings of the loop filter proportional gain β for minimized jitter. However, these analytical results are only valid for the basic BBADPLL architecture as shown in Fig. 4.3 and under some limiting assumptions concerning the amount of DCO jitter $\sigma_{T,\text{DCO}}$ compared to the tuning step T_step [DD08], and some simplifications considering the phase margin of the BBADPLL control loop [ZTL+09]. Therefore in this work a simulation based approach is used to study the behavior of the BBADPLL with respect to the implemented architectures and target applications. It is explained in the following.

4.1.3 Bang-bang ADPLL System Model

The models presented in this section are developed for system analysis of the actual BBADPLL circuit implementations shown in Sec. 4.2 and Sec. 4.3.

4.1.3.1 Basic BBADPLL Model

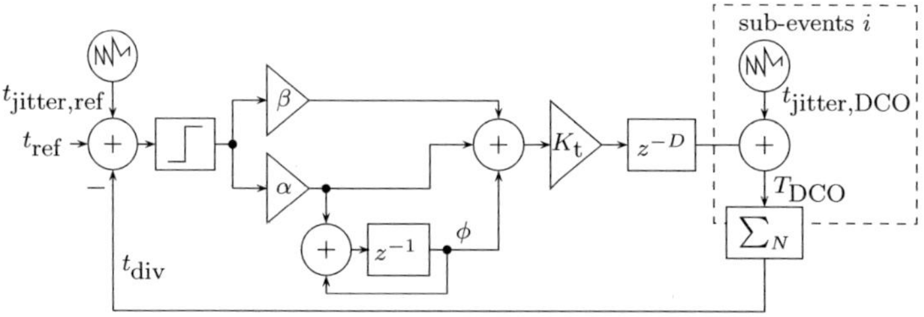

Figure 4.6: BBADPLL model

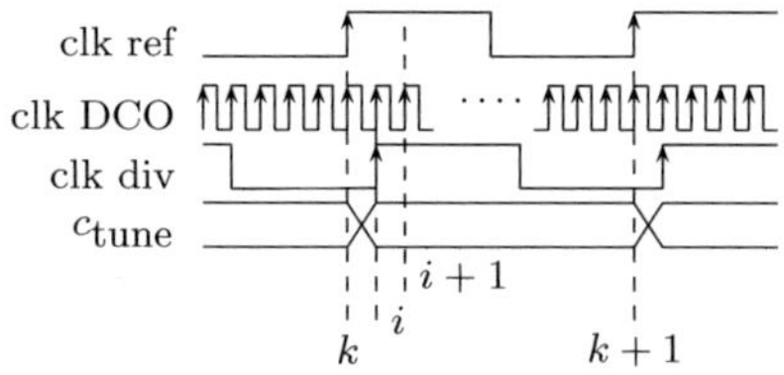

Figure 4.7: Time event indices of BBADPLL model

Fig. 4.6 shows the basic BBADPLL model. It operates on discrete events k for the loop filter. The model equations read

$$t_\text{ref}(k+1) = t_\text{ref}(k) + T_\text{ref} \tag{4.3}$$

$$c(k) = \text{sgn}\left(t_\text{ref}(k) - t_\text{div}(k) + t_\text{jitter,ref}(k)\right) \cdot \alpha + \phi(k) + e \cdot \beta \tag{4.4}$$

$$\phi(k+1) = \phi(k) + \text{sgn}\left(t_\text{ref}(k) - t_\text{div}(k) + t_\text{jitter,ref}(k)\right) \cdot \alpha. \tag{4.5}$$

In the MATLAB implementation the filter value ϕ is limited to the minimum and maximum value, corresponding to the limited value range of the digital hardware implementation.

The DCO is modeled with additional time events with index i. This serves two purposes. First, the DCO periods can be plotted, allowing statistical visualization of the short term jitter metrics (e.g. period jitter) of the ADPLL. Second the inclusion of non-white noise sources (e.g. flicker noise) to the DCO model is enabled, where the time-domain jitter generation method from [HHS08] is used here. The DCO period is calculated by

$$T_{\text{DCO}}(N \cdot (k-1) + i) = T_{\text{offset}} + K_{\text{t}} \cdot c(k-D) + t_{\text{jitter,DCO}}(i). \tag{4.6}$$

The event time of the loop frequency divider is then given by

$$t_{\text{div}}(k+1) = t_{\text{div}}(k) + \sum_{i=N\cdot(k-1)+1}^{N\cdot(k-1)+N} T_{\text{DCO}}(i). \tag{4.7}$$

Eq. 4.2 suggests that there is a maximum ratio α/β allowed for loop stability. As example for $D = 1$ representing one cycle for tuning word calculation, a ratio of $\alpha/\beta < 0.67$ is constrained. Considering only integer values in the basic model the minimum values of the loop filter constants would be $\beta = 2$ for $\alpha = 1$. However, the fact that the proportional path is directly fed to the DCO tuning word suggests that β should be small in order to not increase the tuning step related jitter. For $\beta = 2$ and $\alpha = 1$, the minimum control related jitter would be $\pm 3 \cdot K_{\text{t}}$ in this architecture. In consequence a very small tuning step size, corresponding to a large number of control signal bits would be required to achieve low jitter operation. This provided tough constraints for DCO implementation with respect to the trade-off between tuning range and tuning step size as shown in Sec. 3.1.

4.1.3.2 BBADPLL with DSM

The basic BBADPLL model presented in the previous subsection allows only integer tuning signals. To enable fulfillment of the stability criterion in Eq. 4.2 while maintaining a suitably high tuning step size for simple DCO implementation, fractional DCO tuning can be used [TRF08]. This is realized by a delta-sigma modulator (DSM), as shown in Fig. 4.8. The n LSBs of the tuning signal $c(k)$ are fed to a 1st-order DSM, which produces a 1-bit stream, representing the tuning word LSB as its average value [Höp08]. Thereby the tuning step K_{t} can be increased while

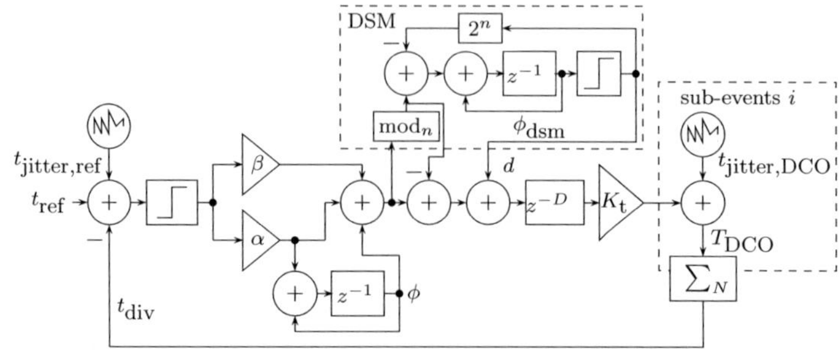

Figure 4.8: BBADPLL model with delta-sigma modulator (DSM)

being able to achieve the stability criterion of α/β. As example, for $n = 5$ LSBs the configuration $K_\text{t} = 1\text{ps}$, $\alpha = 0.0625$ and $\beta = 0.5$ does fulfill Eq. 4.2 for $D = 2$ while achieving a maximum tuning related jitter step of only $\pm 1 \cdot K_\text{t}$. The model equations of the DSM read

$$\phi_{\text{dsm}}(k+1) = \begin{cases} \text{mod}_n(c(k)) + \phi_{\text{dsm}}(k) - 2^n & \text{if } \text{mod}_n(c_{\text{tune}}(k)) + \phi_{\text{dsm}}(k) > 2^n \\ \text{mod}_n(c(k)) + \phi_{\text{dsm}}(k) & \text{else} \end{cases} \tag{4.8}$$

$$d(k+1) = \begin{cases} 1 & \text{if } \text{mod}_n(c(k)) + \phi_{\text{dsm}}(k) > 2^n \\ 0 & \text{else} \end{cases} \tag{4.9}$$

The single bit DSM output d is added to the MSB portion of the tuning word. The DCO control signal of the BBADPLL model with delta-sigma modulation reads

$$c_{\text{DSM}}(k) = c(k) - \text{mod}_n(c_{\text{tune}}(k)) + d(k) \tag{4.10}$$

4.1.3.3 BBADPLL with DSM and Fast Controller Clock

The noise accumulation in the BBADPLL can be decreased when reducing the delay D of the digital filter [DD05, ZTL+09]. Therefore an architecture is proposed here, where the digital filter is running with the divided DCO clock by $N/4$ instead of the reference clock corresponding of a period of T_{DCO}/N when locked. Fig. 4.9 shows the signals of this architecture. The filter operation is divided into four subcycles m. In the first phase $m = 1$ the PFD signal captured and in $m = 2$ the tuning signal c_{tune} is calculated. The rising edge of the divided DCO signal is fed to the

BBPFD in subcycle $m = 4$, where the binary phase decision is made. Thereby the loop filter delay with respect to the reference period is reduced from $D = 2$ for the architectures shown in the previous subsections to $D = 0.5$. The DSM operates on all four subcycles, achieving sufficient scrambling of the fractional part of the DCO tuning signal. The block level model of this BBADPLL architecture is similar to the one shown in Sec. 4.1.3.2, except that the DCO periods are summed up to $N/4$ instead of N for each subcycle m and the state variable assignments are performed in the according m subcycle as explained above.

Note that the signal assignments to the main register components (loop filter accumulator, tuning signal register) are only executed once within the four subcycles, except for the DSM. The power consumption of the digital implementation of this controller architecture therefore is not significantly higher compared to the basic architecture as shown in Sec. 4.1.3.2 when clock gating is employed.

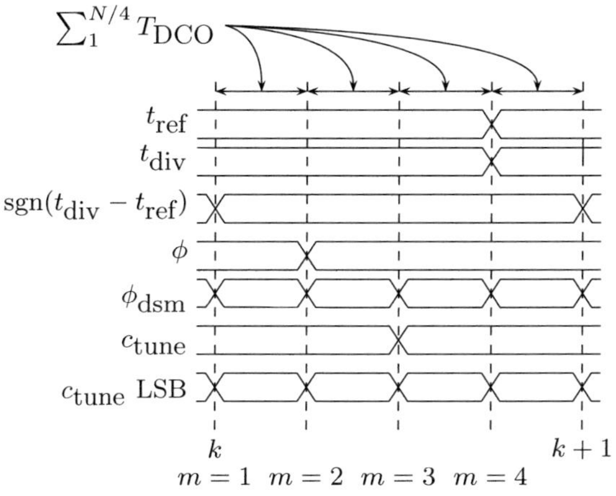

Figure 4.9: Controller timing diagram of the BBADPLL model with DSM and fast controller clock

4.1.4 Bang-bang ADPLL Model Analysis Results

The BBADPLL models are analyzed by numerical simulations using MATLAB. This allows execution of parameter sweeps of the general function

$$(t_{\text{lock}}, \sigma_{t_{\text{ref}}-t_{\text{div}}}, \sigma_{T,\text{acc}}) = F(N, T_{\text{ref}}, \sigma_{\text{T,ref}}, \sigma_{\text{T,DCO}}, \alpha, \beta, D, K_{\text{t}}) \tag{4.11}$$

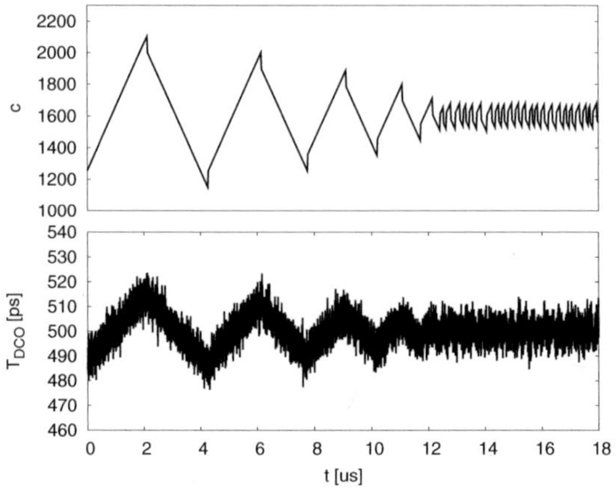

(a) DCO control signal c and period

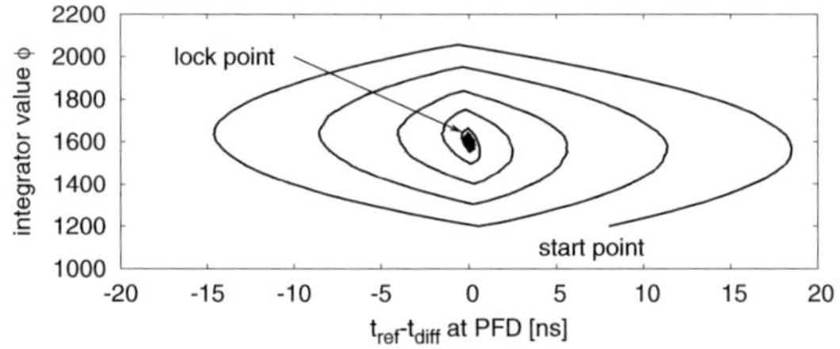

(b) Lock-in trajectory in state space

Figure 4.10: Basic BBADPLL model lock-in simulation, $D = 2$, $K_{\mathrm{t}} = 1/32\mathrm{ps}$, $\alpha = 8$, $\beta = 48$

where F denotes the nonlinear system modeled as described above. Thereby the main ADPLL performances and their parameter sensitivities can be explored. Unless not stated otherwise, analyses are performed with the parameter settings summarized in Tab. 4.1.

Fig. 4.10(a) shows the lock-in behavior of the basic model. When achieving the lock-in region, the DCO control signal shows a periodic oscillation which mainly corresponds to the deterministic orbit of the nonlinear BBADPLL loop, whereas the DCO output period is additionally disturbed by random jitter. Fig. 4.10(b) shows the state space trajectory of this lock-in process.

The lock-in time t_{lock} significantly depends on the loop filter parameters. Fig. 4.11 shows the simulated lock-in time for the BBADPLL (Fig. 4.11(a)) with DSM and

Table 4.1: BBADPLL model parameter settings

T_{ref}	N	T_{offset}	$\sigma_{\text{T,DCO}}$	$f_{\text{flicker,corner}}$	$\sigma_{\text{T,ref}}$
20ns	40	0.45ns	4ps	1MHz	0ps

the version with fast controller clock (Fig. 4.11(b)). First the BBADPLL at lock is simulated to determine the average values $\overline{t_{\text{ref}} - t_{\text{div}}}$, $\overline{\phi}$ and standard deviations $\sigma_{t_{\text{ref}}-t_{\text{div}},\text{lock}}$, $\sigma_{\phi,\text{lock}}$ of the loop filter accumulator value and the PFD input phase difference, respectively. For determination of the lock-in time, the BBADPLL is started from a point $((t_{\text{ref}}-t_{\text{div}})_{\text{start}}, \phi_{\text{start}})$ in its state space and the time is measured until the trajectory reaches the lock-in region defined by $(\overline{t_{\text{ref}} - t_{\text{div}}} \pm \sigma_{t_{\text{ref}}-t_{\text{div}},\text{lock}}, \overline{\phi} \pm \sigma_{\phi,\text{lock}})$. To achieve short lock-in times, large α and β are required.

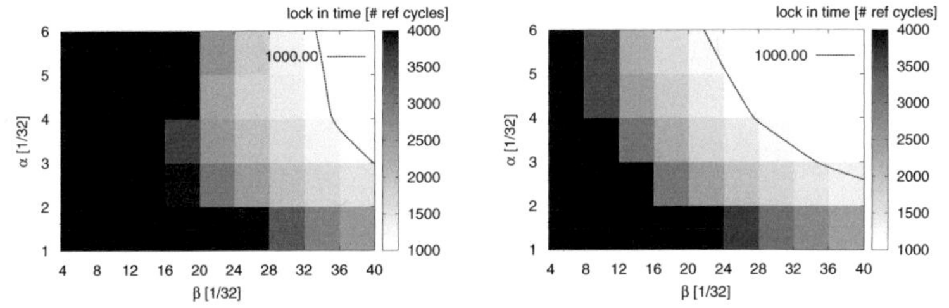

(a) BBADPLL model with DSM ,$D = 2$

(b) BBADPLL model with DSM and fast controller clock

Figure 4.11: Lock-in time t_{lock} simulation of the BBADPLL model, $D = 2$, $K_{\text{t}} = 1\text{ps}$, starting from $\phi_{\text{start}} = 1200$ ($T_{\text{DCO,start}} = 487.5\text{ps}$) and $(t_{\text{ref}} - t_{\text{div}})_{\text{start}} = 8\text{ns}$

Fig. 4.12 shows the dependency of the lock-in time t_{lock} from the initial point in the state space in terms of initial PFD input timing difference and DCO period, from which the BBADPLL is started. The simulations are performed with the BBADPLL with DSM and fast controller clock for different loop filter coefficients as shown in Fig. 4.12(a) and Fig. 4.12(b) respectively. The lock-in time is reduced, if the loop is started closer to its target lock point in the state space. A wider loop filter bandwidth in terms of larger α and β reduces the lock-in time and thereby increases the allowed start region within the state space for a given lock-in time.
Fig. 4.13 shows the simulated absolute BBADPLL jitter as function of the reference clock period jitter $\sigma_{T,\text{ref}}$. For the small reference clock jitter values of $\sigma_{T,\text{ref}} < 9\text{ps}$ and $\sigma_{T,\text{acc},\infty,\text{ref}} < 20\text{ps}$ which are expected for the clock generator application in this work, as shown in App. A.3, the absolute ADPLL jitter does not depend on

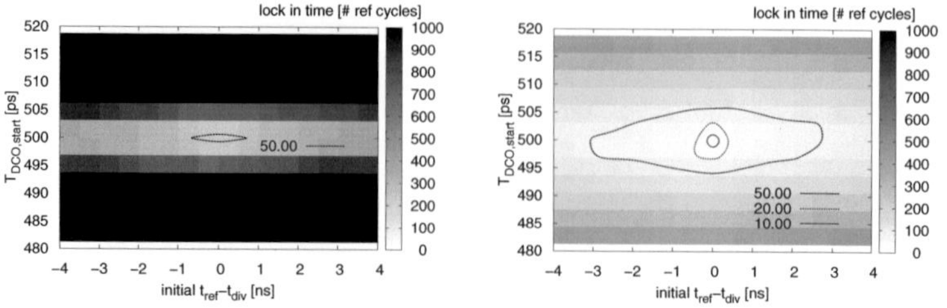

(a) Lock-in plane DSM, fast controller clock, $\alpha = 2/32$, $\beta = 16/32$

(b) Lock-in plane DSM, fast controller clock, $\alpha = 8/32$, $\beta = 64/32$

Figure 4.12: Simulated lock-in time t_{lock} depending on the start condition of the BBADPLL in terms of DCO period and PFD input time difference

the reference clock jitter. Therefore the BBADPLL is operating in a DCO noise dominated regime when using the DCOs shown in Sec. 3.2 and Sec. 3.3 respectively. In the following system model analyses the reference clock jitter is neglected.

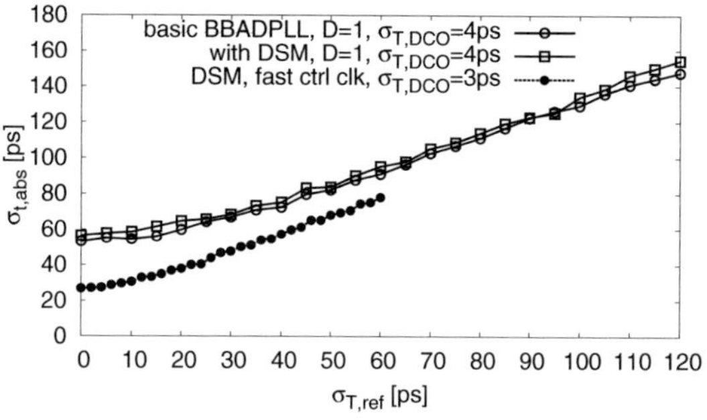

Figure 4.13: Reference clock jitter influence on BBADPLL total jitter

Fig. 4.14 shows the accumulated jitter[1] over n DCO clock cycles for the three BBADPLL model types and different loop delays D. Generally, a small loop delay is desirable for low jitter accumulation. The addition of the DSM for fractional tuning slightly increases the accumulated jitter compared to the basic model with similar parameters. This is caused by the fact that the DSM adds an additional effective

[1] In contrast to [DD05] and [ZTL⁺09] where the standard deviation of the input jitter $\sigma_{t_{\text{ref}}-t_{\text{div}}}$ is calculated, here the accumulated jitter of the DCO over n clock cycles is considered with respect to the jitter tolerance definitions in the DDR2 and DDR3 memory interface standards [JED07, JED09, JED10]. Both measures are related by $\sigma_{T,\text{acc},n\to\infty} = \sqrt{2} \cdot \sigma_{t_{\text{ref}}-t_{\text{div}}}$ as shown in App. A.1

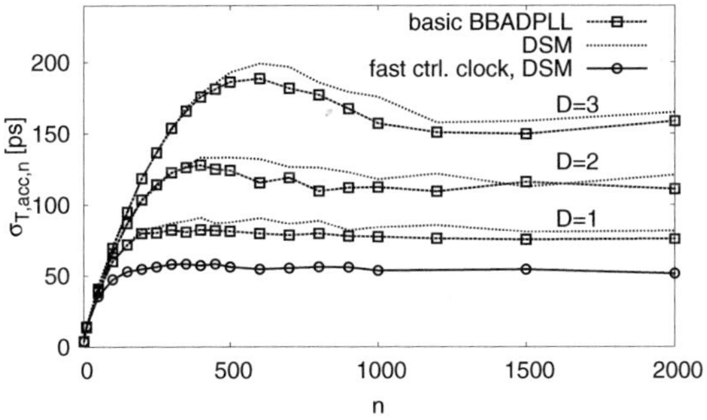

Figure 4.14: Jitter accumulation over n DCO output clock cycles, basic model $K_\mathrm{t} = 1/32\mathrm{ps}$, $\alpha = 2$, $\beta = 16$, DSM Model: $K_\mathrm{t} = 1\mathrm{ps}$, $\alpha = 2/32$, $\beta = 16/32$, DIV clk model $K_\mathrm{t} = 1$, $\alpha = 2/32$, $\beta = 16/32$

delay for the fractional part of the tuning word, because its value is represented by the average of the DSM output pulse sequence. Additionally, the DSM adds quantization noise to the tuning signal which is accumulated in the DCO [Höp08]. As expected, the BBADPLL architecture with the fast controller clock exhibits the least accumulated jitter among the three considered versions.

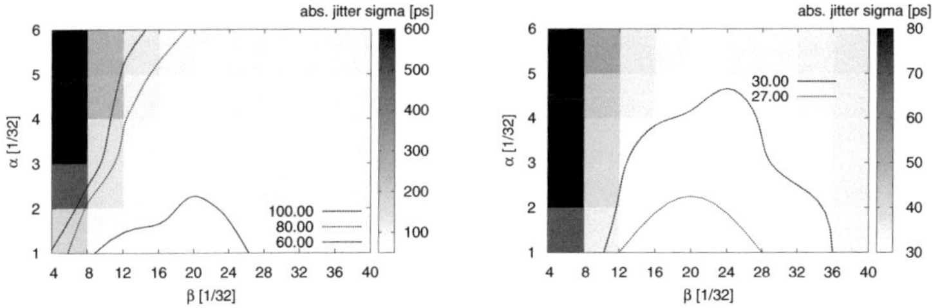

(a) BBADPLL with DSM, $D = 2$

(b) BBADPLL with DSM and fast control clock, $\sigma_{\mathrm{T,DCO}} = 3\mathrm{ps}$

Figure 4.15: Absolute jitter $\sigma_{t,\mathrm{abs}}$, sweeps over α and β,

The jitter accumulation depends on the parameters of the loop filter. Fig. 4.15 shows the accumulated jitter depending on α and β. For a given value of α there exists an optimal value of β, in agreement to the results in [ZTL+09]. For smaller β the total accumulated jitter is dominated by random noise, which is not filtered sufficiently by the loop. For larger β, the deterministic jitter caused by the deterministic orbit of the nonlinear BBADPLL system is increased. Generally, lower α and β improve the jitter performance of the BBADPLL. For the analysis of the BBADPLL model with fast controller clock the DCO period jitter model parameters is changed to $\sigma_{\mathrm{T,DCO}} = 3\mathrm{ps}$ corresponding to the 28nm DCO as presented in Sec. 3.3.

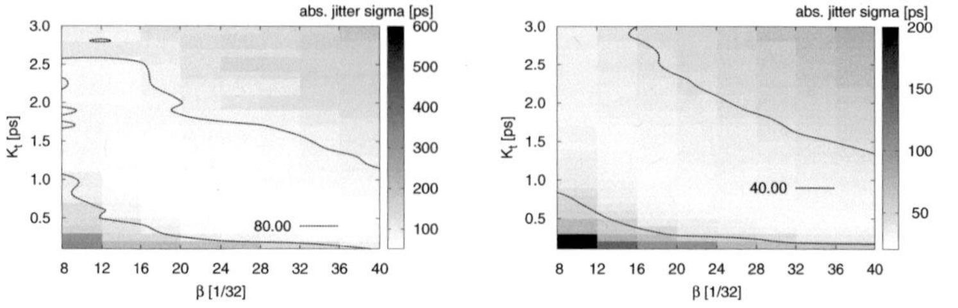

(a) BBADPLL with DSM, $D = 1$, $a = 2/32$

(b) BBADPLL with DSM and fast control clock, $a = 2/32$, $\sigma_{\mathrm{T,DCO}} = 3\mathrm{ps}$

Figure 4.16: Absolute jitter $\sigma_{t,\mathrm{abs}}$ for variations of K_{t} and β

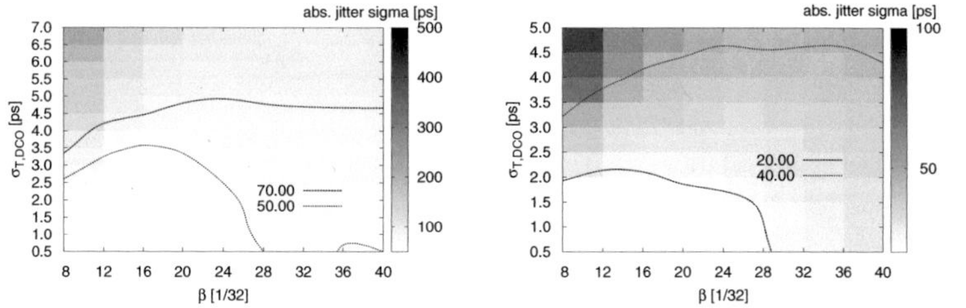

(a) BBADPLL with DSM, $D = 2$, $a = 2/32$, $K_{\mathrm{t}} = 1\mathrm{ps}$

(b) BBADPLL with DSM and fast control clock, $a = 2/32$, $K_{\mathrm{t}} = 1\mathrm{ps}$

Figure 4.17: Absolute jitter $\sigma_{t,\mathrm{abs}}$ for variations of $\sigma_{T,\mathrm{DCO}}$ and β

In contrast to the digitally defined loop filter parameters (D, α, β), the DCO tuning gain K_{t} and period jitter $\sigma_{T,\mathrm{DCO}}$ are linked to the physical DCO circuit, which shows variations with respect to process parameters and operating conditions (e.g. supply voltage, temperature). The closed loop ADPLL must show robustness with respect to these variations. Fig. 4.16 shows the absolute jitter depending on the tuning gain K_{t} and the proportional filter gain β. It shows that low jitter operation can be achieved within a wide range of K_{t} for a given value of β. Therefore an adaptive control of the BBADPLL loop gain [KSK+09], which determines the optimum value of β for minimized jitter during system operation, is not required

for the targeted application. Both DCOs as presented in Sec. 3.2 and Sec. 3.3 are capable of low jitter operation within the BBADPLL architectures as analyzed here, including their variations of K_t. Especially for the 28nm DCO which shows larger variations of K_t resulting from the supply resistance tuning mechanism. Considering its characteristics shown in Fig. 3.34 with K_t in the range from 0.3ps to 2.4ps for the target oscillation period, an optimal value of beta $\approx 24/32$ exists. Fig. 4.17 shows the simulated absolute jitter for variations of $\sigma_{T,\mathrm{DCO}}$ and β. The optimal value of β for minimized absolute jitter shifts to higher values when the DCO period jitter is increased. This can be explained by the fact that the gain of the PFD decreases with increasing DCO period jitter [ZTL+09, PK13]
In summary, the main results from the ADPLL system analysis are:

- Delta-sigma modulation can be applied to implement fractional tuning of the DCO, thereby relaxing the constraints for the minimum tuning step size of the DCO, with acceptable increase of the long term jitter accumulation.
- Reducing the delay of the digital controller by clocking with higher frequencies than the reference clock, significantly reduces the accumulated jitter.
- The BBADPLL loop for the targeted applications shows only slight sensitivity to parameter variations of the custom DCO components. Thus no adaptive loop filter gain control is required.
- For the selection of the loop filter coefficients α and β exists a trade-off between minimum jitter accumulation when locked and the lock-in time t_{lock}.

4.1.5 Fast Lock-in Concepts

The previous analyses have shown the trade off between low output clock jitter when the ADPLL is locked, and the initial lock-in time. For the target application of MPSoC clock generation, both fast lock-in and low jitter are required. This underlines the demand for additional circuit techniques that enable fast lock-in while achieving minimized jitter when locked.

4.1.5.1 Gear Shifting Loop Filter

The digital nature of the loop filter allows to change α and β during operation easily. In principle two or more sets of filter coefficients α and β are used for lock-in, which can be separated in PVT calibration and acquisition phase [SB07], and operational phase tracking phase. These sets of coefficients can be optimized separately for short lock-in time and low jitter, respectively. Thereby the BBADPLL is operating in its closed-loop configuration all times, but includes additional gear shifting functionality to switch the filter coefficients. With the sequential reduction of the loop filter bandwidth during the start up process, the perturbations of the tuning word get smaller, reducing the jitter in the BBADPLL output signal. This technique has been used in [SB07] and [Wag09].

4.1.5.2 DCO Target Period Search

Another approach is to run the BBADPLL during lock-in not in closed loop configuration, where the final DCO tuning value results from the closed loop system, but to determine the target DCO period by other techniques. To speed up the lock-in process, methods for direct search of the DCO period have been applied. In [EMH+09] the actual DCO frequency is measured using a counter based frequency detector. The number of DCO periods within one or more reference cycles is counted and compared to the desired ratio of N, which is the targeted multiplication factor. Based on that a binary search algorithm is used to achieve frequency lock within n reference clock cycles for a n-Bit resolution of the DCO tuning word.

4.1.5.3 Direct Calculation of the Target Period

Fast frequency lock can also be achieved by direct calculation of the DCO target period, when its exact tuning characteristics are known. This calculated value can be directly applied to the DCO. A linear tuning characteristic is shown in Fig. 4.18, which can be written as

$$T_{\mathrm{DCO}} = T_{\mathrm{offset}} + c \cdot K_{\mathrm{t}} \tag{4.12}$$

and is completely defined by T_{offset} and the tuning gain K_{t}.

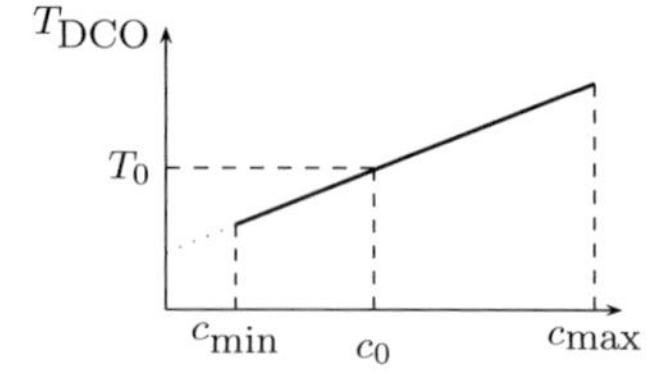

Figure 4.18: Linear DCO tuning characteristics

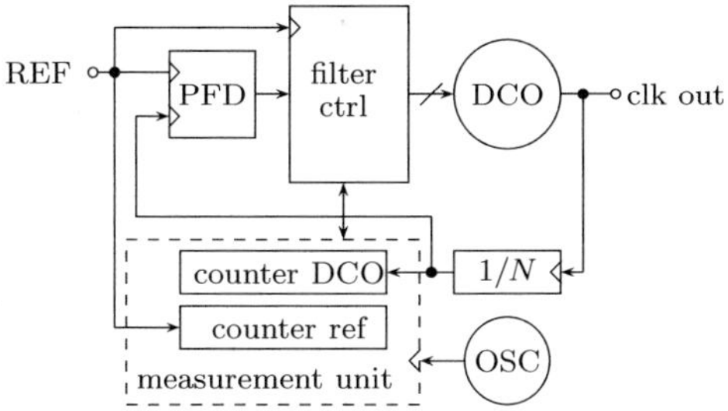

Figure 4.19: ADPLL with time measurement unit and auxiliary oscillator, based on [Wag09]

In [Wag09] an ADPLL architecture with lock-in assist circuits is presented. A simplified schematic is shown in Fig. 4.19. Besides the closed loop BBADPLL controller with loop filter, a time measurement unit is used to measure the DCO tuning characteristics and to directly calculate the target tuning word c_0 for frequency lock. An auxiliary oscillator is used as measurement time base. Its period T_{meas} does not need to be known exactly (as shown below), which minimizes the accuracy constraints for circuit implementation, i.e. a minimalistic ring oscillator with small chip area and low power consumption can be used for this purpose. It is $T_{\text{meas}} < T_{\text{DCO}}$. Since the tuning range of a DCO with switched chain length might be large, this allows effectively to reduce the measurement time for a given relative counting error, compared to a counter which is using the reference clock period (typically $T_{\text{ref}} = 20\text{ns}$) as reference. Additionally, the usage of the auxiliary oscillator can speed up the required lock-in calculations, if the time measurement units are clocked by a multiple of T_{meas}, which is still smaller than T_{ref}. The measurement unit is deactivated during normal BBADPLL operation. Thereby power consumption is not an issue here. Additionally the auxiliary oscillator can also be used to calibrate the DCO tuning characteristics as shown in Sec. 3.1.2.

The periods of the DCO at its maximum and minimum tuning value and the reference period are measured by the counter, resulting in count values y_{min}, y_{max} and y_{ref}, respectively. It is

$$T_{\text{div,max}} = N \cdot T_{\text{DCO,max}} = N \cdot (T_{\text{offset}} + c_{\text{max}} \cdot K_{\text{t}}) = y_{\text{max}} \cdot T_{\text{meas}} \tag{4.13}$$

$$T_{\text{div,min}} = N \cdot T_{\text{DCO,min}} = N \cdot (T_{\text{offset}} + c_{\text{min}} \cdot K_{\text{t}}) = y_{\text{min}} \cdot T_{\text{meas}} \tag{4.14}$$

$$T_{\text{ref}} = N \cdot T_0 = N \cdot (T_{\text{offset}} + c_0 \cdot K_{\text{t}}) = y_{\text{ref}} \cdot T_{\text{meas}}. \tag{4.15}$$

So the target tuning word c_0 can be calculated from the counter measurements by

$$c_0 = c_{\text{min}} + (c_{\text{max}} - c_{\text{min}}) \cdot \frac{y_{\text{ref}} - y_{\text{min}}}{y_{\text{max}} - y_{\text{min}}} \tag{4.16}$$

being independent from T_{meas}. Based on this a new tuning word for a frequency change, corresponding to a change in the BBADPLL frequency divider N, can be calculated and directly be applied to the DCO, which significantly reduces the lock-in time compared to closed loop BBADPLL operation. Details on the algorithms the numerical errors, and the digital hardware implementation of this method can be found in [Wag09].

However, this method can be used for initial login, but may not be suitable for frequency changes during PLL operation because the measurement of y_0 must be performed by detuning the DCO, which prevents to use the BBADPLL output clock for MPSoC system clocking during this phase.

This issue can be circumvented by the ADPLL architecture used in [WSWW10], where the DCO tuning characteristics are monitored during ADPLL operation by two auxiliary DCOs, which replicate the functional DCO. These replicates run at c_{min} and c_{max} respectively and their periods are measured with respect to the reference clock T_{ref} which monitors the linear tuning characteristics as shown in Fig. 4.18 by this two point measurement. The replica DCOs are only used for this purpose and thereby form a significant overhead in terms of chip area.

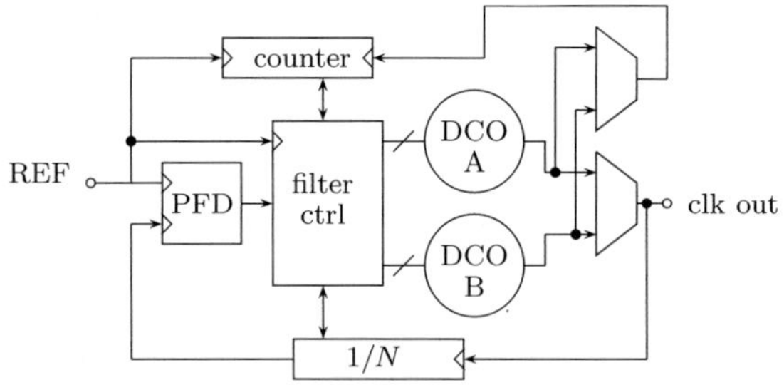

Figure 4.20: Simplified schematic of ADPLL with dual DCOs for on-the-fly calculation of target period, based on [Haa11]

In [Haa11] this technique is improved by using a single DCO replica (DCO B) and the functional DCO (DCO A) for this measurement purpose. This architecture is shown in Fig. 4.20. During lock-in, the linear tuning characteristics of number of periods of DCO B within one reference clock cycle is counted with when DCO B is running at $c_{\min}$ and $c_{\max}$ respectively [2], leading to

$$y_{\max} \cdot T_{\text{ref}} = T_{\text{offset}} + c_{\max} \cdot K_{\text{t}} \tag{4.17}$$

$$y_{\min} \cdot T_{\text{ref}} = T_{\text{offset}} + c_{\min} \cdot K_{\text{t}}. \tag{4.18}$$

For the target tuning word c_0, $y = N$ DCO clock cycles must be within one reference clock. So the c_0 can be calculated by

$$c_0 = c_{\min} + (c_{\max} - c_{\min}) \cdot \frac{N - y_{\min}}{y_{\max} - y_{\min}} \tag{4.19}$$

similar to Eq. 4.16. During ADPLL operation, one of the two DCOs is operating in functional mode with tuning word c_0 and the loop division ratio N, producing the output clock. The replica DCO is running at $c_{\min}$ or $c_{\max}$ depending on the actual c_0, for a maximized difference between the tuning word of functional and replica DCO. The values $y_{\min}$ or $y_{\max}$ are determined. Thereby the target tuning word c_1 for a new frequency corresponding to N_1 can be calculated by

$$c_1 = \begin{cases} c_{\min} + (c_0 - c_{\min}) \cdot \frac{N_1 - y_{\min}}{N - y_{\min}}, & \text{if } N_1 < N \\ c_0 + (c_{\max} - c_0) \cdot \frac{N_1 - N}{y_{\max} - N}, & \text{if } N_1 > N. \end{cases} \tag{4.20}$$

The computation of the target tuning value in Eq. 4.20 represents a linear interpolation. For its algorithmic hardware realization different approaches are possible. To achieve minimum computation time which is desirable for fast ADPLL lock-in, Eq. 4.20 can be directly realized using adders, a multiplier and a divider. However this results in larger logic area. If the computation can be spread over multiple clock cycles, iterative methods (e.g. bisection search) can be applied to realize the target tuning word calculation with minimized hardware effort.

For fast switching, the role of being functional or replica DCO can be flipped between DCO A and DCO B. By this method the DCO tuning characteristics can be tracked during system operation for fast frequency changes for DVFS applications. Details on the algorithms the numerical errors, and the digital hardware implementation of this method can be found in [Haa11].

However, the two techniques presented in this subsection rely on *linear* DCO tuning characteristics, which limits their application to special types of DCOs, like for example the controlled logic chain lengths DCO. (see Sec. 3.1).

[2]This can be done within multiple reference clock periods for increased measurement accuracy

4.1.5.4 Restart in Target Lock Point

As shown in the lock-in time analysis results in Fig. 4.12, fast lock-in can be achieved when (re-)starting the BBADPLL in its target lock point in the state space. This can be feasible when the BBADPLL has been locked once and is disabled for power saving reasons during processor idle phases. A re-lock-in can be achieved by storing digital tuning value (loop filter integrator) and realizing a restart at PFD input phase difference of zero. This principle is used in the novel fast lock-in scheme of the ADPLL in 28nm CMOS technology as described in Sec. 4.3.

4.2 A Compact ADPLL in 65nm CMOS Technology

An ADPLL for local clock generation for MPSoCs is developed and implemented in 65nm CMOS technology. It is used in the testchips "Tommy" and "Atlas" shown in Sec. 2.8. The circuit has been published in [HEH+13], as presented in the following subsections.

4.2.1 Circuit Structure

Fig. 4.21 shows the block level schematic of the ADPLL clock generator. It includes the DCO from Sec. 3.2 which is locked to a nominal period of $T_0 = 500\text{ps}$. From this, the open-loop clock generation circuits as presented in Sec. 5.1 generate the output clocks for the processor core and the high-speed NoC links. The DCO is implemented in a full-custom design style. The frequency divider, BBPFD and the open-loop clock generators are implemented in a custom-digital style using logic cells from a high speed standard cell library (see Sec. 2.7). The ADPLL controller is implemented using a semi-custom synthesis and place&route flow.

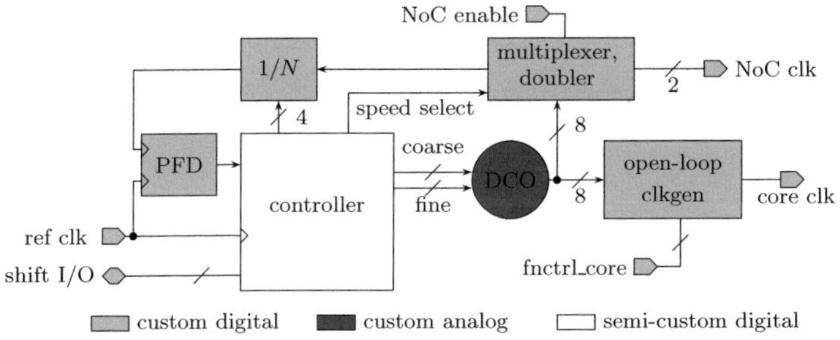

Figure 4.21: ADPLL clock generator block-level schematic, [HEH+13]

4.2.1.1 Frequency Divider

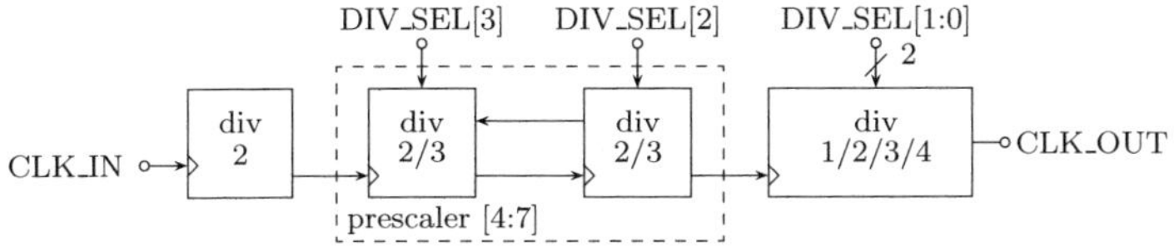

Figure 4.22: Frequency divider by N schematic

An integer-N frequency divider is used in the ADPLL feedback path. Fig. 4.22 shows its block level schematic. It is built up using the topology presented in [VFL+00]. The division ratio is programmable by the select signal (DIV_SEL) which is static during normal operation. Thus no glitch and synchronization issues have to be considered. The first stage divides the DCO output clock by 2 leading to a lower frequency clock for the programmable stages for power saving reasons. The two cascaded divide by 2/3 stages provide frequency division ratios in the range [4 : 7] ([VFL+00]) which are controllable by the two most significant bits (MSBs) of the control signal $c_{\mathrm{DIV\_SEL},3:2}$. The output divider operates synchronously and provides ratios in the range of [1 : 4] and is configured by the two LSBs of the control signal $c_{\mathrm{DIV\_SEL},1:0}$. The total division ratio reads

$$N = 2 \cdot (4 + c_{\mathrm{DIV\_SEL},3:2}) \cdot (1 + c_{\mathrm{DIV\_SEL},1:0}) \tag{4.21}$$

where $c_{\mathrm{DIV\_SEL},3:2}$ and $c_{\mathrm{DIV\_SEL},1:0}$ are the decimal representation of the division ratio select signals. In total 16 different integer division ratios N can be realized in the range from 8 to 56. For nominal operation from a 50MHz reference clock it is $N = 40$ with $c_{\mathrm{DIV\_SEL},3:2} = 1$ and $c_{\mathrm{DIV\_SEL},1:0} = 3$. The programmability provides flexibility to run the ADPLL clock generator at nominal DCO period T_0 from 16 different reference clock frequencies in the range from 35.7MHz to 250MHz.

4.2.1.2 Digital Loop Filter

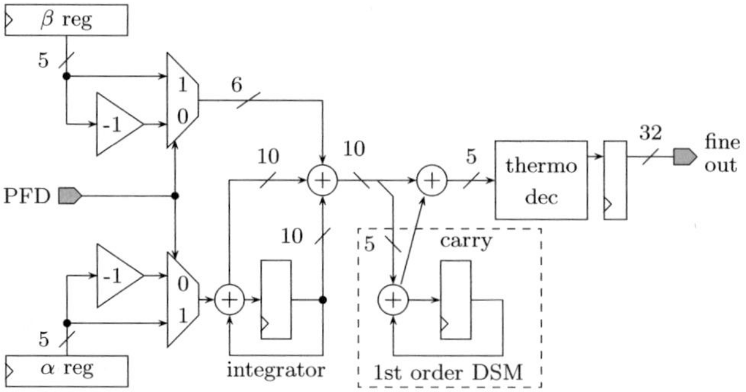

Figure 4.23: Loop filter schematic, all registers clocked with reference clock, bus widths as implemented in the "Tommy" testchip

Fig. 4.23 shows the schematic realization of the digital filter implementing the BBADPLL loop filter architecture as analyzed in Sec. 4.1.2. The tuning signals are represented by fixed-point numbers, where the non-integer part of the tuning signal is fed to the digital DSM realizing fractional DCO fine tuning. This is required because the DCO tuning step K_T is relatively high ($\approx$ 2ps) for the 65nm DCO (see Sec. 3.2) but the filter constants coefficients must fulfill $\beta \gg \alpha$ for stable limit cycles of the nonlinear system as shown in Sec. 4.1.2. As assumed in the system simulation results of the BBADPLL in Sec. 4.1.4 a resolution of 5-bit for the fractional part of the filter signal is chosen here. Especially considering the results of the accumulated jitter simulations in Fig. 4.15 a fractional resolution of α and β of 1/32 is sufficient to allow fine adjustment to the optimal jitter performance while maintaining low hardware complexity. For adaption of the loop filter parameters in the manufactured circuits, α and β can be programmed in the range from 1/32 to 31/32 for optimization of the lock-in time and accumulated jitter performance.

4.2.1.3 Lock Detection

The ADPLL controller must be capable to detect if the phase and frequency of the output signal are locked to the target, before the output clock can be safely fed to the clocked components of the MPSoC. This is indicated by a locked signal which is also used as clock gate enable for the clock generator output.

The lock condition can be detected by the number of BBPFD signal transitions within a given time frame, because the phase error is changing its sign when the bang-bang loop is locked close to the zero phase different point at the PFD input. This behavior is visualized in the system simulation results in Fig. 4.10. Therefore the digital ADPLL controller contains a timer which defines the lock detection window. Within this time frame the number of occurring PFD transitions (1 to 0 and 0 to 1) are counted. The counter value is limited to 7. When this limit is reached within the lock detection time frame, the ADPLL is assumed to be locked. The transition counter is reset to 0 with the beginning of the next lock detection cycle.

4.2.1.4 Bang-bang Phase Frequency Detector

As presented in [HEH$^+$13], Fig. 4.24 shows the BBPFD which is adopted from [TRF08]. It consists of two flip-flops which are clocked by the reference clock and the divided DCO signal respectively. In reset, their outputs are '1'. When one flip-flop is clocked, its output is set to '0'. The following cross-coupled NOR latch ensures that only the first falling signal edge at the flip-flop outputs defines the logic state in the next stages. A meta-stability filter is added to reduce the probability of metastable states when both flip-flops are clocked at nearly the same time (which occurs often when the ADPLL is in lock). The detection result is stored in an output latch built up using cross-coupled NAND gates. When both rising edges have arrived at the flip-flops and the decision has been propagated to the output latch the flip-flops are reset by a self-timed asynchronous reset signal which is generated by a Muller C-element [Lu93]. It sets its output to '1'/'0' if all three input signals are '1'/'0' respectively. Otherwise it holds its output state.

Due to the delay time of the internal asynchronous reset loop, a dead zone exists in the BBPFD timing characteristics. When a signal edge arrives within this time frame t_{dead} it is canceled by the next reset. Thus, the output of the BBPFD indicates the wrong phase and frequency difference. The worst-case dead zone time obtained from simulations in this 65nm implementation is $t_{\text{dead}} < 1\text{ns}$.

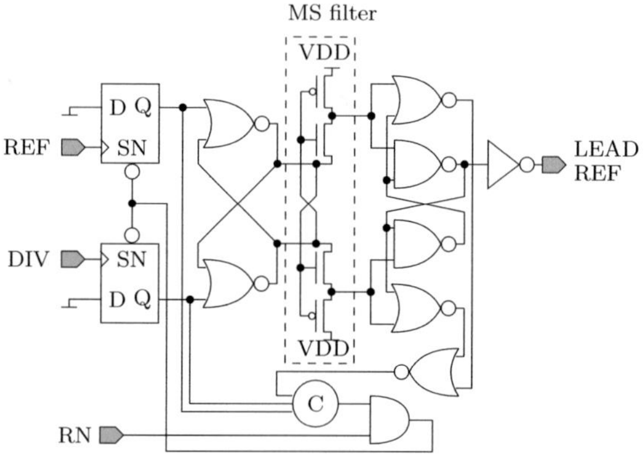

Figure 4.24: Bang-bang PFD, [HEH+13]

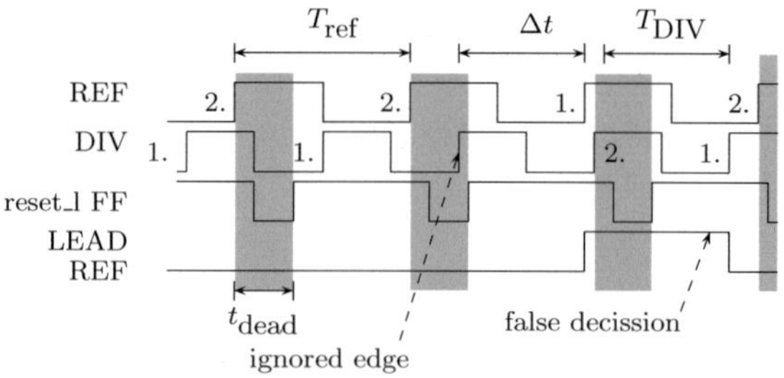

Figure 4.25: BBPFD waveform with wrong frequency decision for $T_{\text{DIV}} < T_{\text{ref}}$

Fig. 4.25 illustrates the conditions for a lost edge that leads to a wrong frequency decision in the BBPFD output. This occurs if the condition

$$T_{\text{ref}} - t_{\text{dead}} < \Delta t < T_{\text{DIV}} \tag{4.22}$$

is fulfilled, for the case $T_{\text{DIV}} < T_{\text{ref}}$, where a similar relation holds for the case $T_{\text{ref}} < T_{\text{DIV}}$ due to the symmetry of the BBPFD circuit. Δt denotes the time difference of the rising clock edges at the input of the BBPFD. From Eq. 4.22 it can be concluded that the probability of a false frequency decision increases when the dead zone t_{dead} is large and the period difference between the reference clock and the divider clock is small (maximum T_{DIV} fulfilling $T_{\text{DIV}} < T_{\text{ref}}$).
The BBPFD can be used for correct detection of the frequency difference between its input signals, if the dead zone does not extend half of the reference period $t_{\text{dead}} < T_{\text{ref}}/2$. Then on average more than 50% of the frequency difference descissions are correct, leading to a DCO frequency adjustment in the right direction [SM90]. With respect to the target reference period of $T_{\text{ref}} = 20\text{ns}$, this condition is fulfilled here.

4.2.2 Coarse Lock-in Mechanism

After starting the ADPLL, a coarse lock sequence is performed. The coarse tune setting c_{coarse} of the DCO is determined such that the frequency lock condition $N \cdot T_{\text{DCO}} \approx T_{\text{ref}}$ is met. Therefore the BBPFD is used as binary frequency detector to compare the reference period T_{ref} with the divided DCO signal of period $N \cdot T_{\text{DCO}}$. This reduces hardware effort compared to additional counter-based frequency detectors, as for example used in [WSWW10] or [CCYL06].
As presented in [HEH+13], a timer defines a time frame $n_{\text{count}} \cdot T_{\text{ref}}$ where the DCO coarse tune signal is kept constant. The output of the BBPFD at the end of this time frame is used as the frequency comparison value.
As explained in the previous subsection (Fig. 4.25), the dead zone behavior of the BBPFD can cause wrong output pulses. Consider a divider signal which is faster than the reference clock, the expected BBPFD output would be zero. If a first rising edge (1.) of DIV occurs within the dead zone, it is ignored and the following rising edge of REF is considered the first one, causing a wrong output value. Due to the beat period of $T_{\text{ref}} - N \cdot T_{\text{DCO}}$ the ignored edge *moves* through the dead zone. The wrong pulse is corrected when the next rising edge of DIV occurs before the rising edge of REF, i.e. after $t_{\text{dead}}/(T_{\text{ref}} - N \cdot T_{\text{DCO}})$ cycles. Thereby the wrong output signal pulse width increases if the ADPLL is closer to its frequency lock condition. To reduce the probability of a wrong frequency comparison, the BBPFD output

is filtered in the controller logic. The filter output is changed to one/zero if n_f consecutive ones/zeros occur at the BBPFD output. Otherwise it keeps its previous value. So the required number of filter cycles to achieve correct frequency difference detection for an DCO period accuracy of ΔT_DCO reads

$$n_\mathrm{f} \geq \frac{t_\mathrm{dead}}{N \cdot \Delta T_\mathrm{DCO}}. \tag{4.23}$$

As example for $N = 40$, $t_\mathrm{dead} = 1\mathrm{ns}$ and $\Delta T_\mathrm{DCO} = 5\mathrm{ps}$ a number of $n_\mathrm{f} \geq 5$ filter cycles is required. The total number of reference cycles to be counted for frequency detection with a maximum resulting error of the DCO period of ΔT_DCO reads

$$n_\mathrm{count} \geq \frac{T_\mathrm{ref}}{N \cdot \Delta T_\mathrm{DCO}} = \frac{T_\mathrm{DCO}}{\Delta T_\mathrm{DCO}}. \tag{4.24}$$

As example for $T_\mathrm{DCO} = 0.5\mathrm{ns}$ and $\Delta T_\mathrm{DCO} = 5\mathrm{ps}$ a number of $n_\mathrm{count} > 100$ reference cycles must be counted. Note that this is a worst-case value assuming that the phase relation between the DCO and the reference clock is completely unknown during frequency detection. Based on this binary frequency detection, linear search or successive approximation [EMH+09] is performed to determine the coarse tune value of the DCO. After this coarse tune phase, fine tune phase lock is achieved by normal closed loop BBADPLL operation.

For illustration of the lock-in behavior Fig. 4.26 shows the digital waveforms of the ADPLL controller during lock-in from RTL simulations. First coarse frequency lock-in is performed by linear search. The BBPFD output contains false pulses, which length increases as the DCO period approaches its target value. Except in the last frequency decision step these pulses can be eliminated using the filter method explained above, leading to a correct frequency lock-in behavior. In the second phase the ADPLL is started in closed loop operation, where phase lock is achieved by fine tuning of the DCO. When the lock-in point with respect to phase and frequency is reached, the BBPFD output signal frequently changes its value, which is detected by the transition counter that finally indicates lock.

4.2.3 Implementation Results

This 65nm ADPLL has been implemented in two testchips "Tommy" and "Atlas" with some slight modifications. The version within "Tommy" as published in [HEH+13] features a simple current DAC based DCO with on-chip current bias and 32 fine tune steps. It employs a basic current bias generator with small supply voltage and temperature dependency, corresponding to the current bias component 0

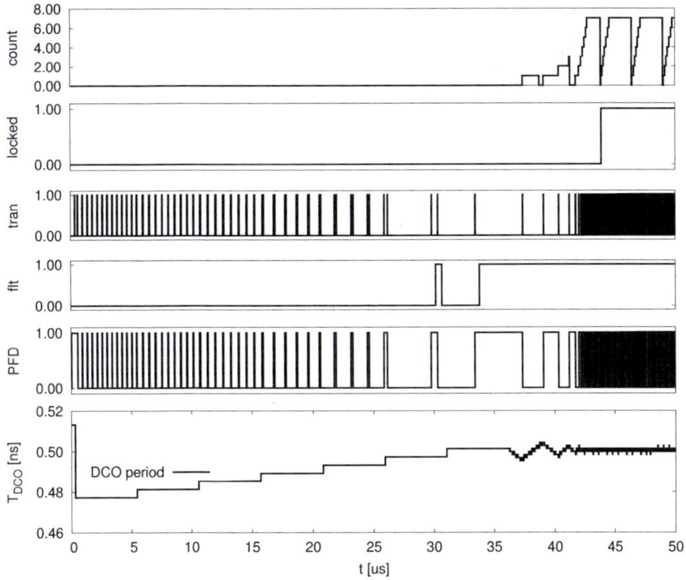

Figure 4.26: ADPLL lock-in RTL simulation results

as presented in Sec. 3.2.4. The enhanced version as realized in the "Atlas" testchip contains the DCO with supply voltage and temperature compensated biasing as presented in Sec. 3.2.4. Additionally the number of fine tune steps has been doubled to 64. Both versions use the open-loop clock generator from Ch. 5 for core clock generation. Fig. 4.27 and Fig. 4.28 show both layouts.

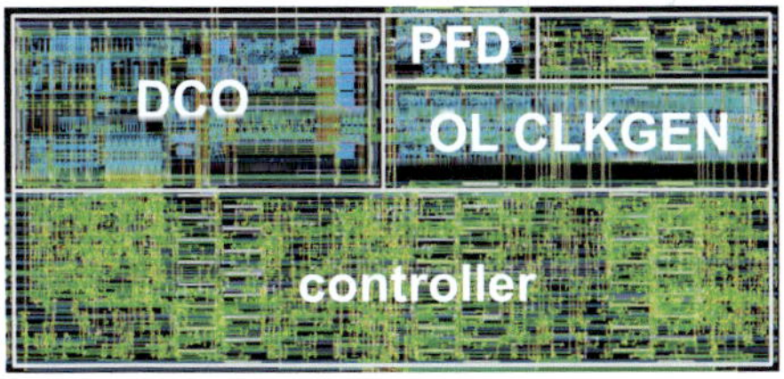

Figure 4.27: Layout of the ADPLL in 65nm CMOS technology, 120μm × 65μm "Tommy" testchip version,

Fig. 4.29 and Fig. 4.30 show the measured period jitter and long term accumulated jitter of the ADPLL realization within the "Tommy" testchip [HEH+13]. The

Figure 4.28: Layout of the ADPLL in 65nm CMOS technology, "Atlas" testchip version, $180\mu m \times 54\mu m$

output clock fulfills the specification for short term jitter (period jitter and cycle-to-cycle jitter) as well as long term accumulated jitter with respect to the DDR2 and DDR3 memory interface clock specification [JED09], [JED07], [JED10], as shown in Fig. 4.31(a) and Tab. 4.2. This shows that the proposed ADPLL can not only be used for efficient core clocking applications but is also capable of driving MPSoC interface components.

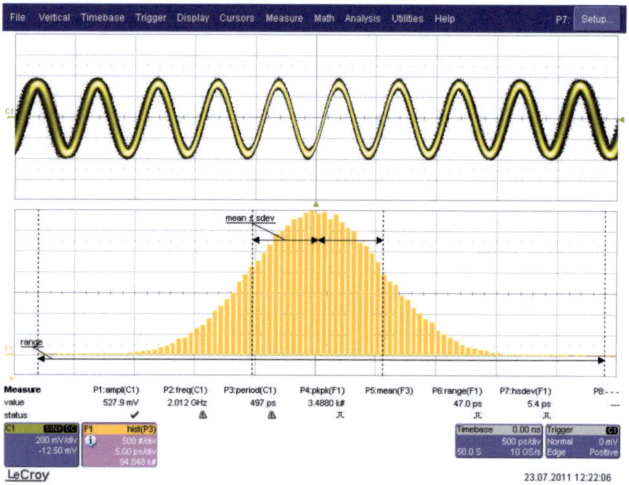

Figure 4.29: Measured period jitter of 2GHz ADPLL clock output, σ_T = 5.4ps, "Tommy" testchip version,

Fig. 4.31(b) shows the measured jitter accumulation for different values of the loop filter coefficients α and β. These results have been obtained from the "Atlas" testchip. The results are in good agreement to the analyses in Sec. 4.1.4. Note that the total accumulated jitter of the ADPLL realization in "Atlas" is higher than in

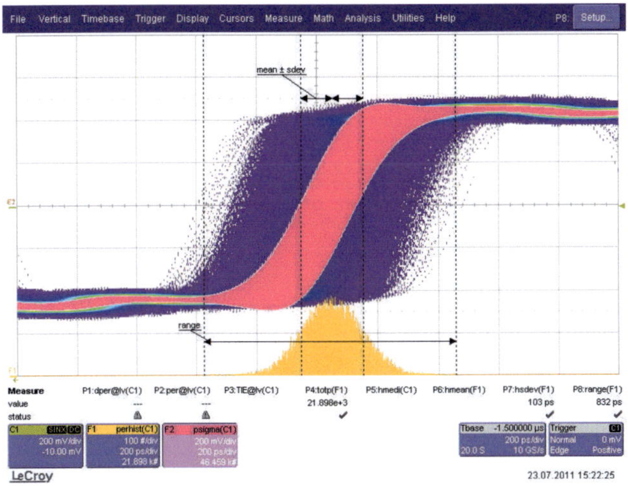

Figure 4.30: Measured long term accumulated jitter histogram, "Tommy" testchip version, $\sigma_{T,\mathrm{acc},\infty} \approx 103$ps

Table 4.2: DDR2/DDR3 period jitter specification, "Tommy" testchip

speed		spec (±pp)	meas (σ)	peak/σ
667	$j_{T,\mathrm{pp}}$ [ps]	125 (100)	20	6.2 (5.0)
667	$j_{CC,\mathrm{pp}}$ [ps]	250 (200)	19	13.1 (10.5)
800	$j_{T,\mathrm{pp}}$ [ps]	100 (90)	17	5.9 (5.3)
800	$j_{CC,\mathrm{pp}}$ [ps]	200 (180)	15	13.3 (12.0)
1066	$j_{T,\mathrm{pp}}$ [ps]	90 (80)	14	6.4 (5.7)
1066	$j_{CC,\mathrm{pp}}$ [ps]	180 (160)	13	13.8 (12.3)
1333	$j_{T,\mathrm{pp}}$ [ps]	80 (70)	11	7.3 (6.4)
1333	$j_{CC,\mathrm{pp}}$ [ps]	160 (140)	9	17.8 (15.5)

"Tommy" due to the additional current source components in the active DCO bias circuit. However additional noise does not translate into period jitter because its bandwidth is significantly smaller than the nominal DCO oscillation period. But it is accumulated over several DCO cycles and therefore leads to an increased long term jitter. In future re-designs the noise of the bias circuit can be reduced by additional filter capacitances in the static bias circuit parts or a general increase of the bias circuit power consumption. Tab. 4.3 summarizes the main performances of the MPSoC clock generator in 65nm CMOS technology employing the open-loop clock generator presented in Ch. 5.

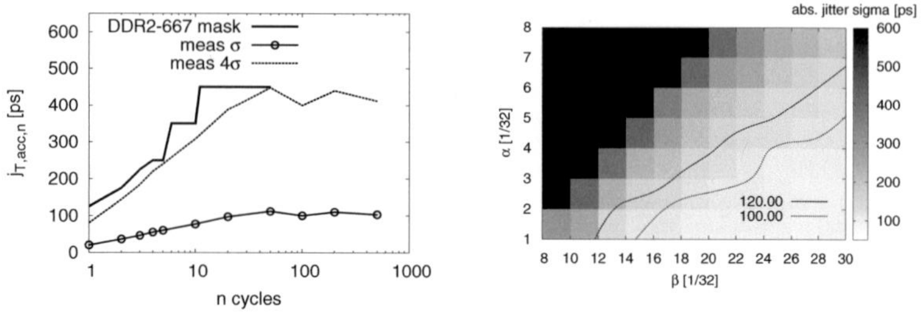

(a) accumulated jitter with respect to DDR2 specification, "Tommy" testchip

(b) absolute jitter, estimated by $\sigma_{t,\mathrm{abs}} = 1/\sqrt{2} \cdot \sigma_{T,\mathrm{acc},\infty}$, $V_{\mathrm{DD}} = 1.2\mathrm{V}$, $K_{\mathrm{t}} \approx 1.0\mathrm{ps}$, "Atlas" testchip

Figure 4.31: Measured jitter of 65nm ADPLL, room temperature

Table 4.3: Typical 65nm clock generator performances

f_{DCO} [GHz]	2	
N	8 to 56	main divider
f_{NoC} [GHz]	2 or 4	
f_{core} [MHz]	83 to 666	33 frequencies
lock time	$< 100\mu$s	
core clock change	0μs	
$V_{\mathrm{DD,DCO}}$ [V]	1.2	analog supply
$V_{\mathrm{DD,core}}$ [V]	1.2	core supply
P_{DCO} [mW]	1.90 / 2.05	w./w.o. doubler at 1.2V
P_{ctrl} [mW]	0.16	at 1.2V
P_{olclkg} [mW]	0.60 to 1.65	at 1.2V
DCO $\sigma_{T,\mathrm{DCO}}$ [ps]	5.4 / 52.0	rms / pp
core $\sigma_{T,\mathrm{core}}$	$< 0.8\% T_{\mathrm{core}}$	
accumulated jitter $\sigma_{T,\mathrm{acc},\infty}$ [ps]	103	
area [μm^2]	7800	

4.3 A Fast Locking ADPLL in 28nm CMOS Technology

The ADPLL clock generator concept from Sec. 4.2 has been implemented in GLOBALFOUNDRIES 28nm CMOS technology. It is used as versatile clock generator within the "Cool28SoC" testchip as presented in Sec. 2.8. Modifications include an improved clocking concept for the digital ADPLL controller and novel fast-lock-in functionality.

4.3.1 Circuit Structure

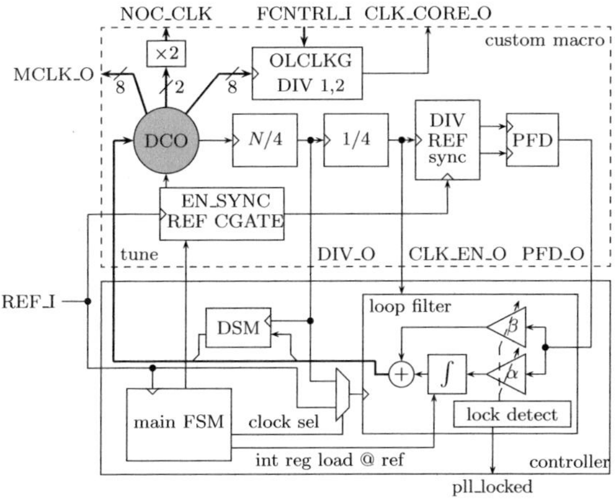

Figure 4.32: Block level schematic of the ADPLL in 28nm CMOS

Fig. 4.32 shows the block level schematic of the ADPLL in 28nm CMOS technology. Its custom design part includes the multi phase DCO from Sec. 3.3 providing output clocks for frequency doubling and core clock generation by open-loop methods as presented in Sec. 5.1. Further it includes the ADPLL loop frequency divider with ratio N and a synchronizer block to realize fast lock-in as presented in Sec. 4.3.2. The loop divider provides two output clocks with periods of $T_{\mathrm{DCO}} \cdot N/4$ and $T_{\mathrm{DCO}} \cdot N$, respectively, which are employed to realize a digital loop filter clocking scheme with fast divider clock as shown in Fig. 4.9. The clock timing diagram of the controller is shown in Fig. 4.33. This enables a fast filter response which results in lower accumulated jitter within the bang-bang loop as shown in Sec. 4.1.4.

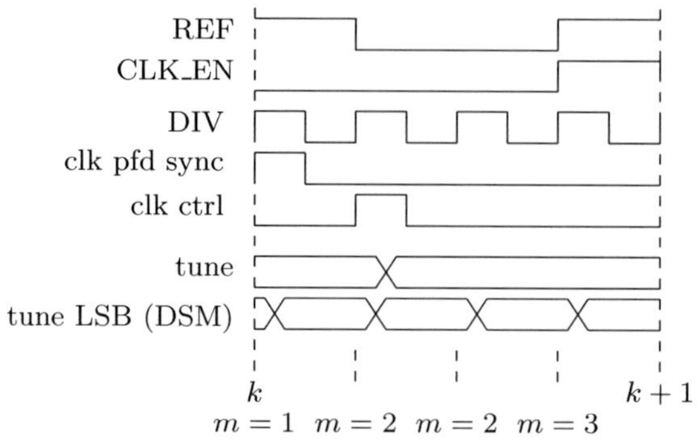

Figure 4.33: Controller timing diagram

The digital part of this ADPLL contains a main finite state machine (FSM) and a loop filter with lock detection similar to the one presented in Sec. 4.2. The state sequence for fast lock-in and operation is shown in Fig. 4.34. The main FSM controls the lock-in sequence. Therefore it runs with the reference clock being available when the DCO is disabled. When running the coarse lock-in sequence as presented in Sec. 4.3.2, the main FSM can directly control the integrator register of the loop filter to set the tuning word c_{tune} of the DCO. Therefore the loop filter can be operated with the reference clock as well. Before changing to closed loop ADPLL operation the main FSM switches the clock source of the loop filter to the divider clock. At this time the DCO is disabled, which prevents the generation of glitches when switching the clocks. The DCO is enabled by the main FSM and the loop filter operates in closed loop with the clocking scheme from Fig. 4.9. The lock-in detector monitors the transitions of the PFD signal to indicate lock of the ADPLL. A different set of loop filter coefficients α and β is used for LOCK_IN and LOCKED state to realize a gear-shifting filter as described in Sec. 4.1.5.1 for speed up of the fine lock-in process.

4.3.2 Fast Phase-lock Architecture

The 28nm ADPLL realization features mechanisms for fast lock-in. As presented in Sec. 4.1 the lock-in of a BBADPLL includes phase and frequency lock. Both conditions

$$\Delta t_{\text{PFD}} = t_{\text{ref}} - t_{\text{div}} = 0 \tag{4.25}$$

$$N \cdot T_{\text{DCO}} = T_{\text{ref}} \tag{4.26}$$

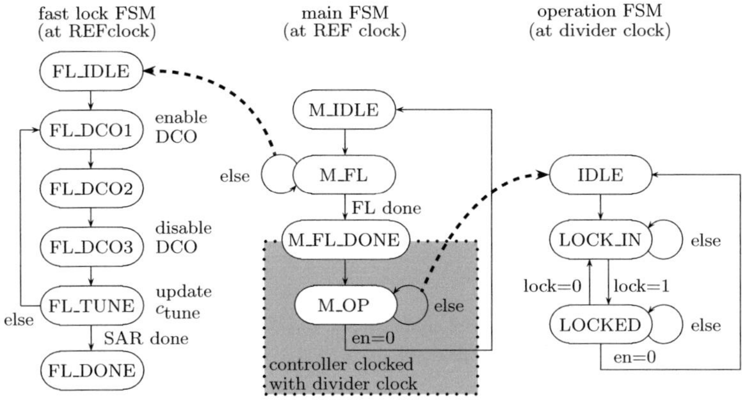

Figure 4.34: ADPLL controller state sequence

must be fulfilled. For minimized lock-in time, the BBADPLL would have to be started in the target lock point of its state-space plane as described in Sec. 4.1.5.4. In contrast to the conventional ADPLL as shown in Fig. 4.35(a), where the phase lock condition is a result from closed loop operation, an active single-shot phase synchronizer is proposed in this work as shown in Fig. 4.35(b). By means of a configurable delay chain with signal edge detection capability it resets the phase difference at the PFD input to zero with the first reference clock edge after starting the ADPLL. After this it keeps its delays from the divider output and the reference input to the PFD static during closed loop operation to allow phase tracking for jitter compensation by the ADPLL. This concept is applicable in this work where the ADPLL is used as *frequency* multiplier for GALS MPSoC clocking. Therefore no defined phase relation between the clock generator output signals and the reference clock is required.

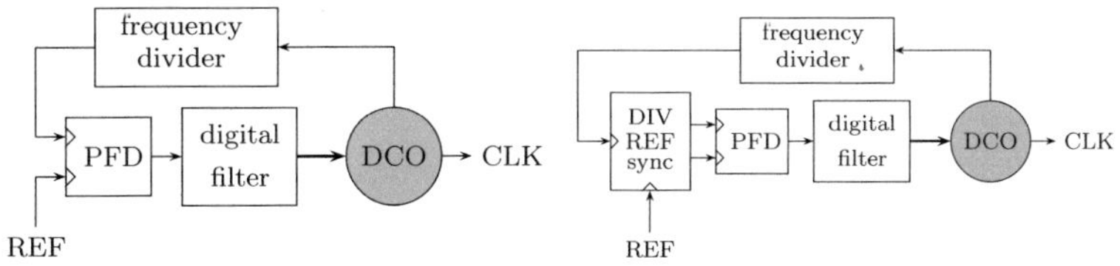

(a) conventional

(b) with single-shot phase synchronizer

Figure 4.35: ADPLL architectures

A similar fast lock-in architecture based on phase synchronization between the

DCO and the reference clock signal has been proposed previously in [WSWW10]. However, the additional delays of asynchronous loop frequency divider stages are not considered there. In contrast, this work provides a versatile phase synchronization solution which is applicable to a wide range of ADPLL architectures by addition of the single-shot phase synchronizer.

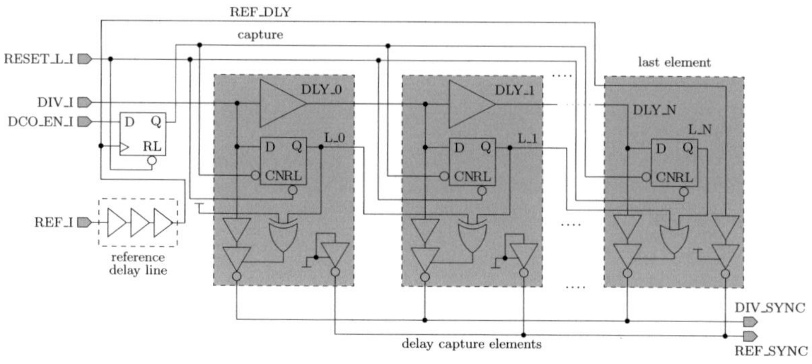

Figure 4.36: Single-shot phase synchronizer schematic, bypass circuits for disabling the synchronization are not shown

Fig. 4.36 shows the schematic of the single-shot phase synchronizer. It synchronizes the rising edges of the divider output clock and the reference clock. Therefore the divider clock (DIV_I) is fed to a main delay line. After each delay element the signal can be selected and fed to the output tri-state bus. The selection bit is stored in a D-latch, which is transparent with the low clock phase. They are transparent after reset. During the capture cycle, the latches store the value of the delayed divider signal in the main delay line. In the acquisition cycle the output of the capture signal flip-flop changes from 0 to 1 with the rising edge of the reference clock. This disables the latches in the main delay line. Thereby the position in the main delay line, where the state of the latches changes from 1 to zero between two adjacent stages, denotes the point where the rising edges of the divider signal and the reference clock occurred at the same time. It is detected by XOR gates (OR in the last stage), which enable the output tri-state drivers for this particular stage. The un-gated reference clock is fed to the output to a replica tri-state stage for symmetry reasons. In the following clock cycles all latches are intransparent because the gated reference clock is static 1. The phase synchronizers remains in its captured state. All delay variations between the divider signal and the reference clock that occur during ADPLL operation are fed directly to the PFD input. Thereby the phase synchronizer does not disturb the operation of the closed-loop ADPLL.

Fig. 4.37 shows the post-layout simulation results of the single-shot phase synchronizer for typical, worst-case and best-case corners. The nominal timing error $|t_{\text{offset,sync}}| < 80$ps for an input skew within ± 200ps. The phase synchronizer needs to compensate the timing variability from synchronous DCO start-up to the first rising clock edges at the PFD input where this variability is mainly caused by the delay through the frequency divider stages. Therefore the compensation range of ± 200ps is sufficient for the application in this work. However, this range can be extended by addition of more delay stages at cost of power consumption and chip area.

Fig. 4.38 visualizes the signal timings for enabling the ADPLL with single-shot phase synchronization. First, the loop dividers are released from reset and the DCO is enabled with the rising edge of the reference clock, triggered by the main FSM of the ADPLL. This results in a first rising edge at the divider output pulse clock (CLK_EN_O), which is fed to the delay line of the single-shot phase synchronizer, where it is synchronized to a delayed copy of the reference clock.

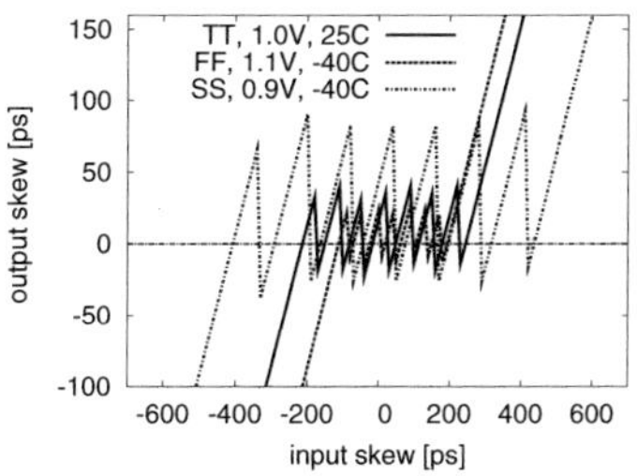

Figure 4.37: Single-shot phase synchronizer simulation results, post layout

It is proposed to use the phase synchronization effectively for binary frequency detection for fast frequency lock to fulfill the condition of Eq. 4.26. When the first signal edge after the reset is synchronized to the reference edge by the single-shot phase synchronizer, the relative position of the following edges is a measure for the relative period of the reference clock and the divided DCO clock, as illustrated in Fig. 4.39. Therefore the PFD combined with the proposed synchronizer can effectively be used as binary frequency detector with low hardware effort.

When the phase is synchronized the initial phase difference is t_{offset}, which is expected to be ideally zero. It includes the offset time of the PFD itself and the remaining timing difference after the enable synchronizer logic $t_{\text{offset,sync}}$. The tim-

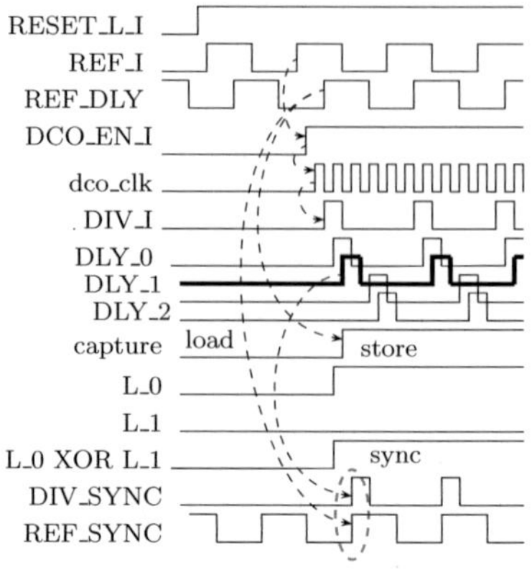

Figure 4.38: Enable timing for fast phase lock restart

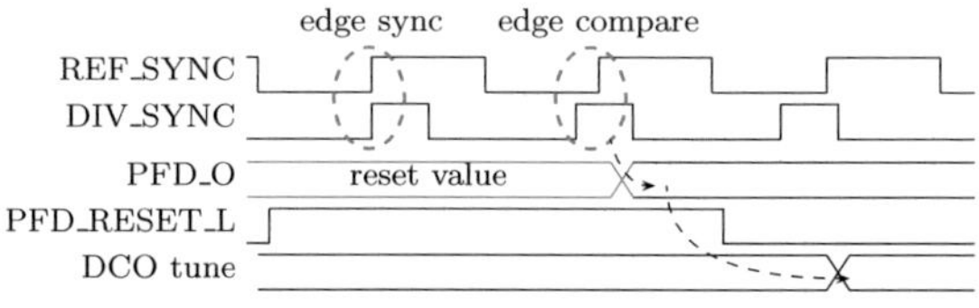

Figure 4.39: PFD Waveform for binary frequency detection

ing difference of the following n-th edge at the PFD input can be calculated by

$$\Delta t_{\text{PFD}}(n) = t_{\text{offset}} + \sum_{i=1}^{i+N\cdot n} T_{\text{DCO},i} - \sum_{j=1}^{n} T_{\text{ref},j} \tag{4.27}$$

where N is the loop divider ratio. From Eq. 4.27 the average value and the standard deviation of Δt_{PFD} is

$$\overline{\Delta t_{\text{PFD}}}(n) = t_{\text{offset}} + n \cdot (N \cdot T_{\text{DCO}} - T_{\text{ref}}) \tag{4.28}$$

$$\sigma^2_{t_{\text{PFD}}}(n) = n \cdot (N \cdot \sigma^2_{T_{\text{DCO}}} + \sigma^2_{T_{\text{ref}}}) \tag{4.29}$$

To determine the number of signal edges n for a required resolution for the DCO period ΔT_{DCO}, first the systematic offset t_{offset} of $\overline{\Delta t_{\text{PFD}}}(n)$ is considered. From Eq. 4.28 it can be concluded that the systematic deviation of the DCO period is

$$|\Delta T_{\text{DCO,systematic}}| = \frac{|t_{\text{offset}}|}{n \cdot N}, \tag{4.30}$$

for constant T_{DCO} and T_{ref} (neglecting jitter). Considering k-sigma accuracy for determination of the DCO period with respect to random jitter, it is

$$k \cdot \sigma_{t_{\text{PFD}}} < n \cdot N \cdot (|\Delta T_{\text{DCO,max}}| - |\Delta T_{\text{DCO,systematic}}|) \tag{4.31}$$

$$k \cdot \sqrt{n \cdot (N \cdot \sigma^2_{T_{\text{DCO}}} + \sigma^2_{T_{\text{ref}}})} < n \cdot N \cdot |\Delta T_{\text{DCO,max}}| - |t_{\text{offset}}| \tag{4.32}$$

where $\Delta T_{\text{DCO,max}}$ is the maximum allowed deviation from the DCO period. So it is

$$|\Delta T_{\text{DCO,max}}| = \frac{1}{n \cdot N} \cdot \left(|t_{\text{offset}}| + k \cdot \sqrt{n} \cdot \sqrt{N \cdot \sigma^2_{T_{\text{DCO}}} + \sigma^2_{T_{\text{ref}}}} \right). \tag{4.33}$$

Solving Eq. 4.33 for n leads to the minimum number of measurement edges for a maximum DCO period variation within k-sigma accuracy

$$n_{\min} = \left(\frac{1}{2N|\Delta T_{\text{DCO,max}}|} \cdot \left(k\sigma_{\text{jitter}} + \sqrt{k^2\sigma^2_{\text{jitter}} + 2N|\Delta T_{\text{DCO,max}}| \cdot |t_{\text{offset}}|} \right) \right)^2 \tag{4.34}$$

with $\sigma^2_{\text{jitter}} = N \cdot \sigma^2_{T_{\text{DCO}}} + \sigma^2_{T_{\text{ref}}}$. Fig. 4.40 visualizes the results from Eq. 4.33. There exists a trade-off between lock-in time and accuracy of the binary frequency search. For selection of a suitable value of n, the BBADPLL analysis results from Sec. 4.1.4 (Fig. 4.12(b)) are used. From this, a DCO period resolution of $|\Delta T_{\text{DCO,max}}| < 3$ps is constrained, with the target of achieving ADPLL lock within 20 reference clock cycles for the results in Fig. 4.12(b). So a value of $n = 2$ is chosen here.

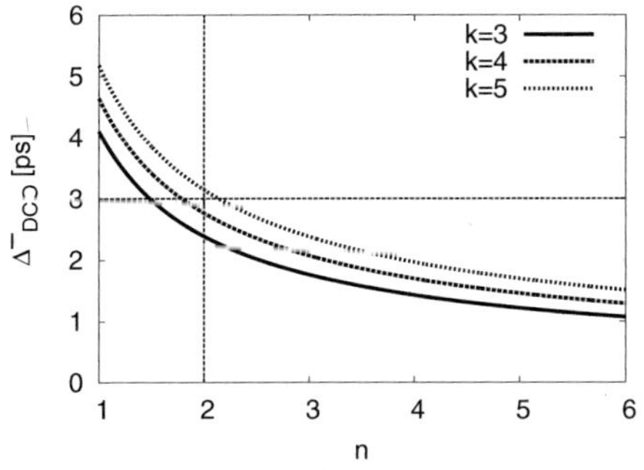

Figure 4.40: Maximum DCO period estimation error versus number of PFD frequency compare cycles n, $\sigma_{T_{\text{DCO}}} = 3$ps, $\sigma_{T_{\text{ref}}} = 10$ps, $N = 40$, $t_{\text{offset}} = 100$ps and different statistical safety margins k

This binary frequency detection scheme is used for frequency lock of the ADPLL.

The by N divided DCO period is compared to the reference period T_{ref}. The output of the PFD indicates the sign of the period difference. Based on this a successive approximation algorithm [EMH+09], where the tuning word setting of m-Bit accuracy is determined within m measurements. Additionally, the number of successive approximation steps can be reduced if the condition $t_{\mathrm{step}} < \sigma_{T_{\mathrm{DCO}}}$ is fulfilled, because the result of the fast lock-in sequence will be in the noise floor of the DCO jitter. Therefore lock-in time can be reduced without significant impact on jitter performance.

Fig. 4.41 shows the waveforms from RTL simulation of the ADPLL. First the binary frequency search is performed in 9 binary frequency detection cycles corresponding to the 10-bit tuning word. Then the closed loop ADPLL is activated with a single-shot phase synchronization of the first rising edges at the PFD input. The lock detection operates similar to the one presented in Sec. 4.2 by counting the number PFD transitions in a given time frame. When not yet locked a wider ADPLL filter bandwidth is achieved by shifting (logic shift left) the loop filter coefficients α and β in a gear shifting filter scheme as presented in Sec. 4.1.5.1. When lock is detected they are set back to their nominal value.

4.3.3 Implementation Results

The ADPLL has been implemented in GLOBALFOUNDRIES 28nm CMOS technology using the flow as presented in Sec. 2.7. Fig. 4.42 shows its compact layout. The circuit is integrated in the "Cool28SoC" testchip shown in Sec. 2.8.

Fig. 4.43 shows the measured 2GHz output waveform with period jitter histogram of the closed-loop ADPLL operation. Fig. 4.44 shows an oscilloscope waveform of the long term accumulated jitter measurement (31ps rms jitter over 2000 accumulated cycles). The detailed sweep over the accumulated jitter over the number of clock cycles is shown in Fig. 4.45(a). Also this ADPLL implementation fulfills the clock jitter specifications for DDR2 and DDR3 memory interfaces.

As presented in Sec. 4.1.4, the accumulated output jitter depends on the loop filter coefficients α and β and the DCO tuning gain K_{t}. Fig. 4.45(b) shows the measured accumulated jitter with sweeps over α and β. These measurements are in good agreement to the model analysis results shown in Fig. 4.15(b) and Fig. 4.16(b).

The fast lock-in functionality is measured by capturing the DCO period over time using the digital sampling oscilloscope. The time 0 of the measured waveforms corresponds to the first rising clock edge of the DCO signal after turning on the ADPLL. Fig. 4.46(a) shows the measured lock-in waveform when both binary frequency search and phase lock with gear shifted loop filter coefficients (by factor

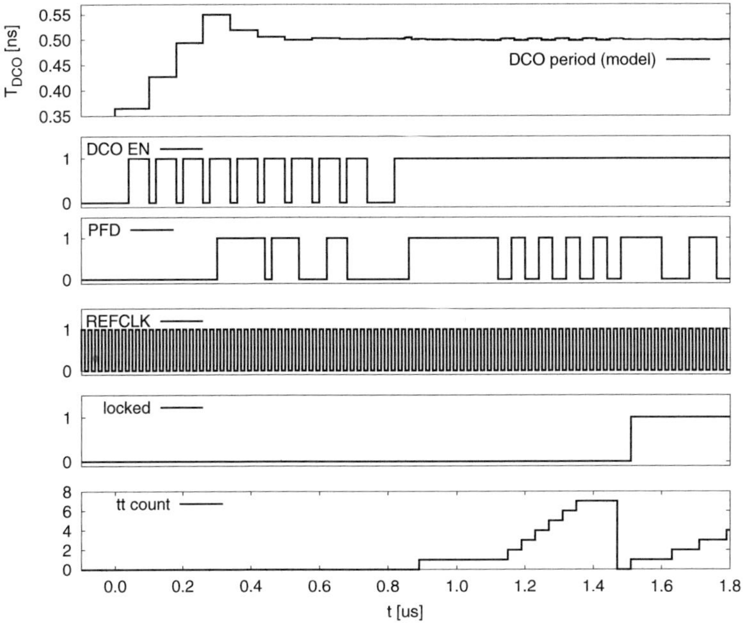

Figure 4.41: Fast lock-in ADPLL RTL simulation results

8) is enabled. Here the lock condition is valid after $0.8\mu s$ and detected after $1.5\mu s$. Fig 4.46(b) shows the lock-in wave if the lock signal flag is directly asserted after the binary frequency search and no gear shifting filter operation is used. It can be seen that the target oscillation period of $T_0 = 500ps$ is directly hit by the frequency search after $0.8\mu s$ and does not change during closed loop operation due to the initial phase synchronization.

Considering this, the proposed phase synchronization method allows to restart the ADPLL immediately, if the previous integrator value of the loop filter (corresponding to the DCO period) is stored. This can be valid in an application scenario where the ADPLL is switched off when the MPSoC core is in idle state if a pausible GALS clocking scheme [KFG+11] is applied. When restarting after a short time (short with respect to temperature drifts in the system), the frequency tuning value is still valid. Fig. 4.47(a) shows this measured instantaneous restart capability. For comparison Fig. 4.47(b) show the restart waveforms *without* phase synchronization. Although started at the correct frequency, the phase lock needs to be acquired by

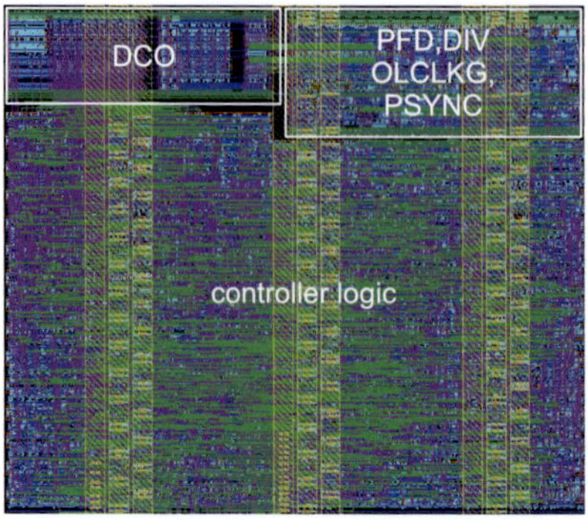

Figure 4.42: Layout of the ADPLL in 28nm CMOS technology, 52μm × 45μm

shifting the DCO period. This leads to increased jitter and prevents using the ADPLL output signal for clocking applications right after startup.
Tab. 4.4 summarizes the main performances of the 28nm ADPLL using the open-loop clock generator as presented in Ch. 5.1 for core clock generation. The power of the single shot phase synchronizer has been measured by differential measurement with and without activated synchronizer. The power overhead is only $\approx 10\mu$W.

Table 4.4: Typical 28nm clock generator performances

f_{DCO} [GHz]	2	
N	8 to 56	main divider
f_{NoC} [GHz]	2	
f_{core} [MHz]	83 to 666	33 frequencies
lock time	$< 2\mu$s	
core clock change	0μs	
$V_{\mathrm{DD,DCO}}$ [V]	1.0	analog supply
$V_{\mathrm{DD,core}}$ [V]	1.0	core supply
P_{DCO} [mW]	0.36	at 1.0V
P_{ctrl} [mW]	0.13	at 1.0V
P_{olclkg} [mW]	0.2	at 83MHz, 1.0V
$P_{\mathrm{phasesync}}$ [mW]	0.01	at 83MHz, 1.0V
DCO σ_T [ps]	3.0	estimated by measurements, see App. A.3
core $\sigma_{T,\mathrm{core}}$	$< 0.8\% T_{\mathrm{core}}$	
accumulated jitter $\sigma_{T,\mathrm{acc},\infty}$ [ps]	30	
area [μm^2]	0.00234mm^2	

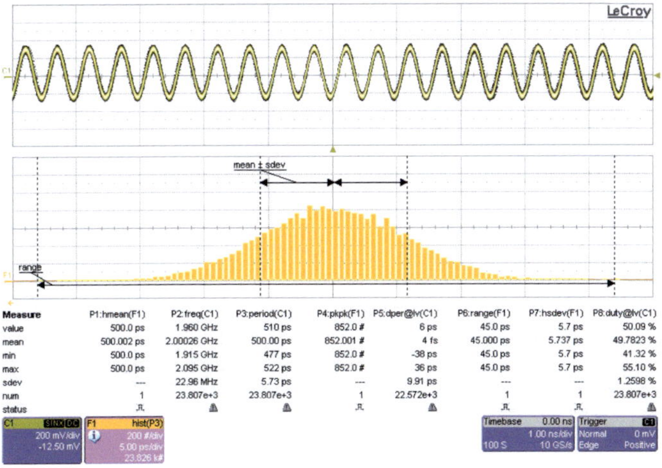

Figure 4.43: Measured 28nm ADPLL signal and period jitter histogram at 2GHz

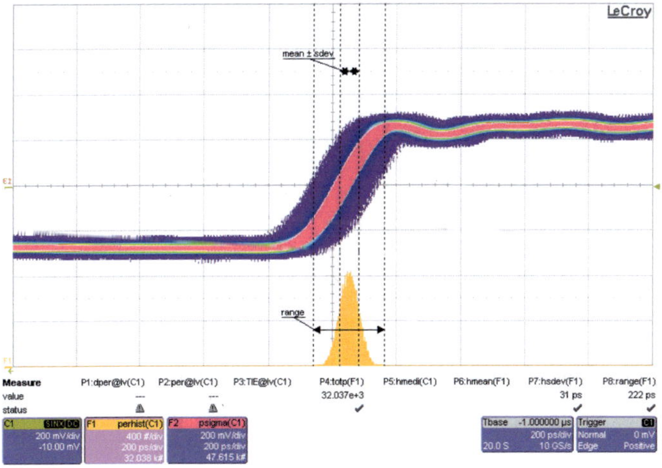

Figure 4.44: Measured 28nm ADPLL long term jitter histogram

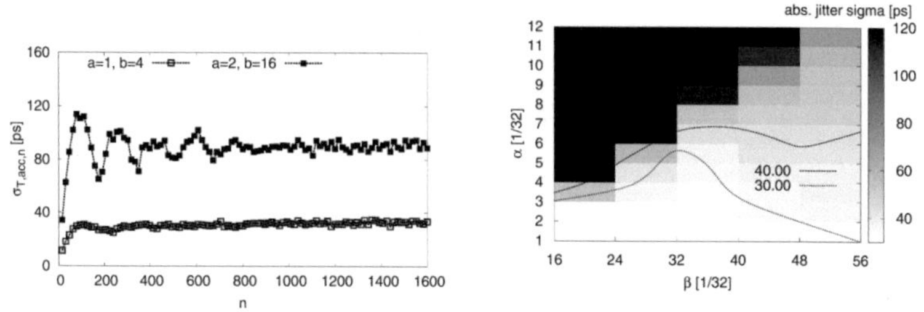

(a) jitter accumulation sweep

(b) absolute jitter, estimated by $\sigma_{t,\mathrm{abs}} = 1/\sqrt{2} \cdot \sigma_{T,\mathrm{acc},\infty}$, $V_{\mathrm{DD}} = 1.0\mathrm{V}$, $K_{\mathrm{t}} \approx 0.6\mathrm{ps}$, room temperature

Figure 4.45: Measured jitter of 28nm ADPLL

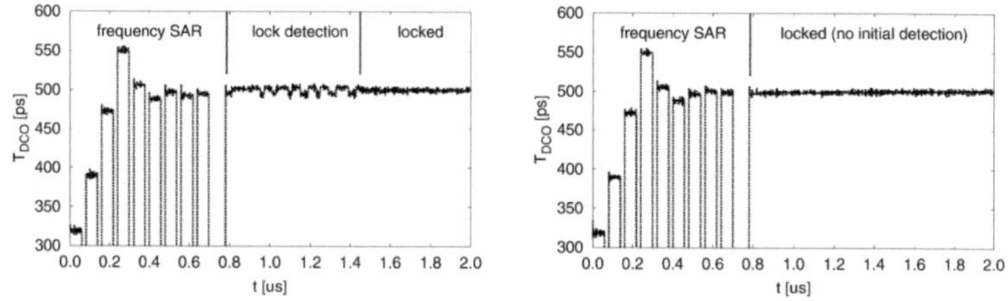

(a) fast lock-in with lock detection and filter bandwidth adaption by factor 8

(b) fast lock-in without initial lock detection, no filter bandwidth adaption

Figure 4.46: Measured 28nm ADPLL lock-in waveform,

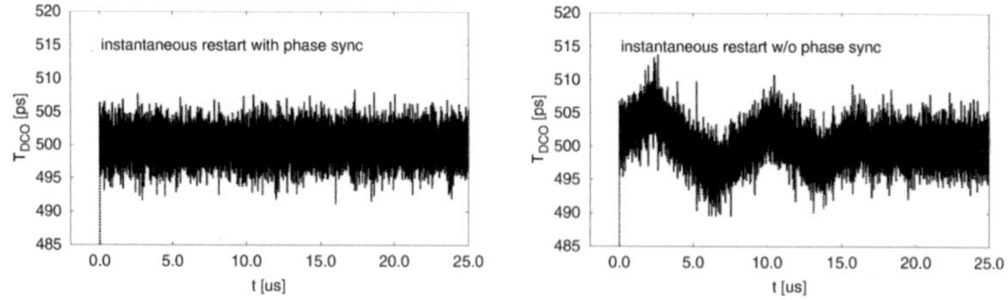

(a) with phase synchronization

(b) without phase synchronization

Figure 4.47: Measured 28nm ADPLL instantaneous restart waveform

4.4 Design Comparison

Table 4.5: Performance Comparison of PLL clock generators in sub-100nm CMOS technologies

Ref	tech [nm]	type	$f_{\min}$ [MHz]	$f_{\max}$ [MHz]	σ_T [ps]	at f [MHz]	P [mW]	FOM_J	A [mm^2]
[Xiu07]	90	FA [1]	2	250	9.0	148	10.0	5.1	0.1512
[YIE+11]	90	ADPLL	700	3500	1.6	2500	1.6	4.0	0.3600
[YCYL12]	90	ADPLL	180	530	n.a.	480	0.466	5.0	0.0086
[TRF08]	65	ADPLL	500	8000	0.7	4000	33.6	4.8	0.0300
[YYG08]	65	PLL	1600	3200	3.1	1600	1.62	4.4	0.0400
[HMY10]	65	ADPLL	3.5	1800	2.6	1600	220.0	6.4	0.5600
[HL09]	65	PLL	900	1000	3.1	900	10.0	4.9	0.1400
[CL10]	65	PLL	850	1100	4.5	1000	8.4	5.2	0.3200
[CL09]	65	PLL	1200	1800	5.4	1500	17.0	5.9	0.2000
[CSM10]	65	ADPLL	600	800	22.0	400	3.2	5.8	0.0270
[GNDD10]	65	ADPLL	190	4270	1.4	3000	11.8	4.8	0.0400
[RTE+08]	45	ADPLL	840	13300	1.1	3800	16.5	4.9	0.0280
[LOK+12]	22	ADPLL	600	3600	n.a.	n.a.	18.4	n.a.	0.0296
this	65	ADPLL	83	4000	5.4	2000	2.7 [2]	5.2	0.0097 (0.0078)[3]
this	28	ADPLL	83	2000	3.0	2000	0.64 [4]	4.1	0.00234

[1]Flying Adder Frequency Synthesizer
[2]at $f_{\mathrm{core}} = 83$MHz and $f_{\mathrm{NoC}} = 2$GHz
[3]with and without DCO bias compensation
[4]at $f_{\mathrm{core}} = 83$MHz and $f_{\mathrm{NoC}} = 2$GHz

The performances of recently published ADPLL clock generators in sub-100nm CMOS technologies are summarized in Tab. 4.5 and compared to the ADPLL implementations in this work. The standard deviation of the period jitter σ_T is chosen as main criterion for output clock quality. Commonly the DCO frequency is much higher than the reference frequency of the ADPLL (clock frequency multiplication), uncorrelated thermal noise as main contributor to DCO period jitter is not filtered by the closed ADPLL loop. Therefore in contrast to [GKGN09] only the DCO noise is considered for benchmarking, assuming that it is mainly caused by thermal noise and that the DCO is the main contributor to power in a ADPLL optimized for low *period jitter*. So the figure of merit is defined as

$$\mathrm{FOM}_J = \log_{10}\left(\frac{\sigma_T^2}{\mathrm{ps}^2} \cdot \frac{f}{\mathrm{MHz}} \cdot \frac{P}{\mathrm{mW}}\right). \tag{4.35}$$

The ADPLL realizations of this work with their low power consumption achieve a similar FOM_J compared to previously published results. They provide a wide range of output frequencies using the open-loop clock generation methods presented in

Ch. 5. The chip area of these all-digital clock generator realizations is very small because no area consuming analog loop filter components are used here. This allows to use the ADPLL clock generators from this work for per-core instantiation within power managed MPSoCs.

4.5 Summary

ADPLLs are suitable architectures for local clock generation in GALS MPSoCs. Especially the BBADPLL topology with simple binary phase frequency detector enables compact circuit realizations. A numerical BBADPLL model has been developed for system analysis and architecture exploration. It has been shown that BBADPLLs are capable to control the DCO to achieve sufficient long term jitter clock quality. A modified controller architecture with faster clocking additionally reduces the accumulated jitter.

An ultra-compact ADPLL clock generator has been implemented in 65nm CMOS technology and has been successfully verified by testchip measurements. It can generate low jitter clocks that meet the DDR2/DDR3 memory interface clock specifications. Its extremely low chip area of 0.0097mm^2 and low power consumption of typically $< 3\text{mW}$ makes it ideally suited for per-core instantiation within GALS MPSoCs.

A second compact ADPLL has been implemented in 28nm CMOS technology with chip area of 0.00234mm^2 and power consumption of $< 0.64\text{mW}$. In addition to the 65nm version this features a controller with fast clocking scheme for less jitter accumulation and a novel fast lock-in scheme based on single-shot phase synchronization at the PFD input. Its behavior has been analyzed mathematically. It allows fast binary frequency detection for frequency lock-in by binary search and can realize instantaneous phase lock at ADPLL start-up. When restarting the ADPLL from power down instantaneous lock-in can be achieved. The circuit functionality has been successfully verified by testchip measurements.

5 Open-loop Clock Generation

The ADPLLs presented in the previous chapter can provide a fixed frequency multi-phase clock (at nominal 2GHz frequency). To address the demand for ultra-fast DVFS core frequency changes as motivated in Ch. 2 open-loop clock generator techniques are developed in this chapter. They allow to generate a wide range of output frequencies from a multi-phase input signal by means of phase rotation and frequency division and enable instantaneous frequency changes. Also frequency multiplication for special purpose high-speed clock generation is addressed here. The sensitivity of the open-loop architectures with respect to phase mismatch due to process variations is theoretically analyzed. Since the local clock generators must be seamlessly integrated into the MPSoC implementation flow, a timing model and constraints concept is shown here.

5.1 Open-loop Clock Generation

Heterogeneous MPSoCs require a wide range of clock frequencies for processor cores, I/O interfaces (e.g. high-speed FPGA interfaces, DDR2/3 memory interfaces) or high-speed on-chip communication links (see. Sec. 2.4). General clock quality requirements like low jitter or 50% duty cycle can easyly be fulfilled by closed loop ADPLLs as shown in the previous sections. But these closed loop clock generators exhibit some mayor drawbacks with respect to their application in heterogeneous MPSoCs.

- Standard integer-N [SSS05] PLLs have a limited number of output frequencies, that are defined by the reference clock frequency and the available frequency division ratios in the loop. Fractional-N PLLs with multi-modulus loop dividers [HSN09] can be a solution to this but impose additional challenges for closed loop PLL design (e.g. filtering of fractional spurs [Höp08])

- The oscillator frequency in the closed loop PLL is the highest frequency in the system. Thereby, high frequency clocks in the GHz range require high-speed oscillators which challenges their circuit implementation.

- The closed-loop PLL approach provides only one output frequency at the same time. This prevents sharing of one closed loop PLL for different clocking applications within the MPSoC, e.g. simultaneous clocking of a high-speed NoC link at some GHz and the processor core with some 100MHz.

- Changes in the output frequency that are required for power management techniques [KFA+07] like DVFS (Sec. 2.2.1) and AVFS (Sec. 2.2.2) during system operation require re-lock of the closed loop PLL which is time consuming [Fah05]. Within this re-lock phase no defined clock frequency is available at the PLL output. Core operation must be paused during re-lock.

To overcome these drawbacks, open-loop clock generators are developed in this work. Based on a fixed frequency signal which is provided by a closed-loop PLL (as shown in Fig. 5.1) output clocks are generated by frequency division (lower frequencies) or multiplication (higher frequencies). These open-loop clock generators use multiple clock phases of the closed-loop PLL to generate the output signal. Thus the output frequency can be tuned independently from the PLL, which provides a wider range of available output frequencies and allows fast changes of the output frequency without re-locking the closed loop PLL. Additionally, multiple open-loop clock generators can be used with one PLL.

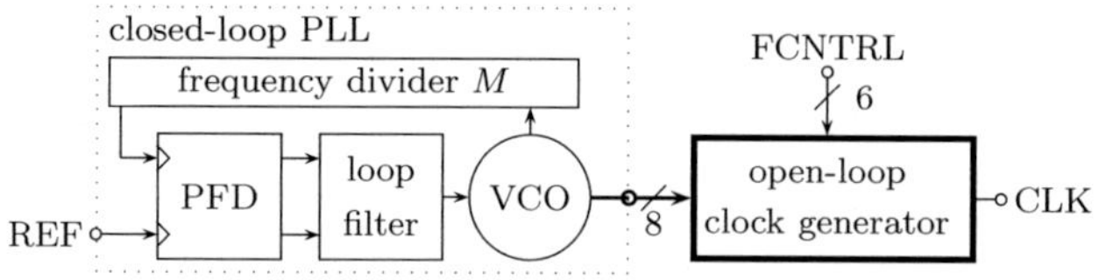

Figure 5.1: Clock generator based on closed loop PLL with multi-phase output and open-loop output clock generator, [HHES11]

The open-loop approach has previously been used in both DLL and PLL based clock generators ([LCL09], [KKK+06], [XY03],[Xiu07], [SLH+10]). PLL based open-loop frequency synthesizers with fine output frequency resolution are realized in the flying adder synthesizers ([XY03],[Xiu07], [SLH+10]), where a multi-phase oscillator signal is used for output clock generation by multiplexing and frequency division. As motivated in Sec. 2.6 and presented in the previous sections, an ADPLL clock generator with multi-phase outputs is used in this work. Both, frequency multiplication and division are employed to generate a wide range of output clocks. These techniques and circuits are presented in the following.

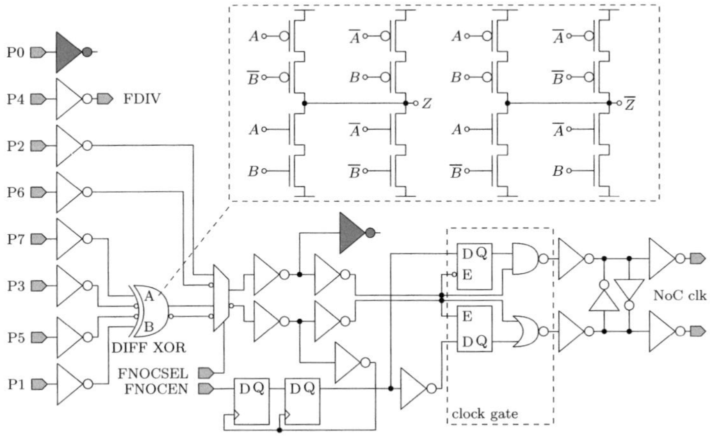

Figure 5.2: NoC clock generator with open-loop frequency doubler

5.1.1 Clock Frequency Multiplication

For clocking of NoC links for on-chip communication the 8 output clock phases of the ADPLL clock generators presented in Sec. 4.2 and Sec. 4.3 are combined to perform frequency doubling. Fig. 5.2 shows the schematic of the NoC clock generator. A differential XOR gate [EFS96] is used to generate a differential clock signal with period $T_0/2$ from 4 phases of the ADPLL output period of T_0. The waveform is shown in Fig. 5.3. Two of the remaining clock phases can be directly multiplexed to the output to run the NoC transmitter with T_0 in a low-speed mode. One single ended output is used for the frequency divider of the closed-loop ADPLL and the last one remains unused. Dummy inverter loads ensure symmetry of the clock phases. Differential clock gating circuits with enable synchronization are added to allow glitch-free activation and de-activation of the NoC clocks. The circuit is implemented using gates from a high-speed digital standard cell library (see Sec. 2.7). Fig. 5.4 shows the layout.

5.1.2 Clock Frequency Division

As presented in [HHES11], several approaches of frequency division with a wide range of division ratios and high resolution have been published previously. The flying adder frequency synthesizer [XY03] employs a phase multiplexer which selects one out of multiple phase clocks. The multiplexer select signal is generated by arithmetic circuitry (adder) operating at the input frequency of the divider (or

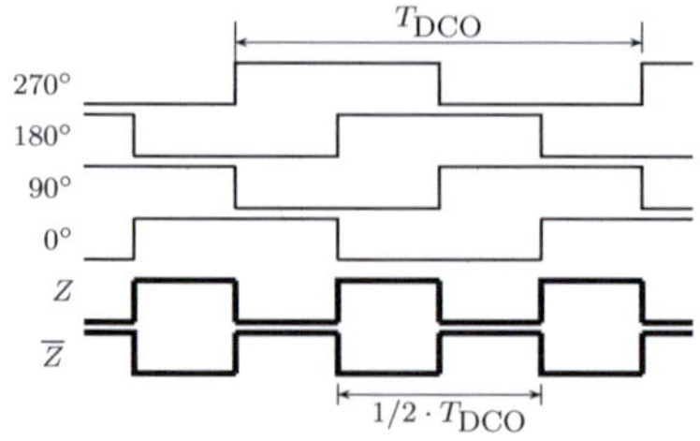

Figure 5.3: Waveform of XOR based frequency doubling

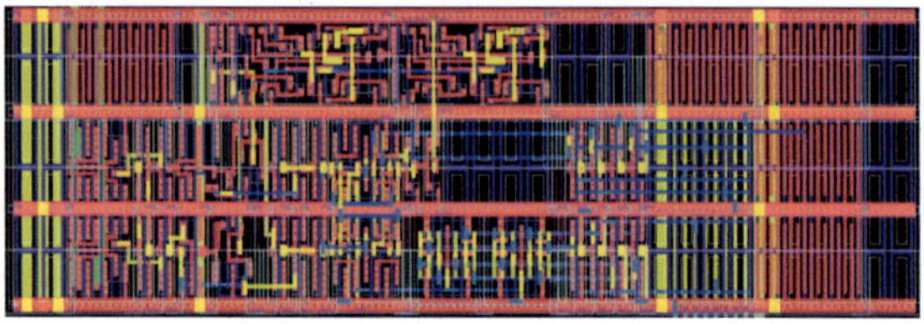

Figure 5.4: NoC clock generator layout, 65nm CMOS, 27μm × 8.4μm

half of it). This limits the maximum operating frequency especially for a higher number of clock phases, but frequency division ratios of less than 1 can be achieved by this technique. Flying adders have been applied successfully for wide range clock generation [Xiu07], [SLH$^+$10]. By reducing the number of clock phases the operating frequency of the flying adder can be increased [XY05]. Phase switching frequency dividers [CS96] are usually employed for fractional-N frequency synthesis in closed loop PLLs [HSN09]. They contain a phase multiplexer with select signals being generated at a fraction of the input frequency. This allows much higher input frequencies but the provided sub-integer division ratios are usually much higher than 1 [WZQW09]. A general theory of phase switching frequency dividers has been presented in [Flo08]. [CC08] shows a glitch free frequency synthesizer based on phase switching. [PMP09] proposes a phase switching divider architecture where the phases of the input signals are calibrated during operation in order to reduce fractional spurs in the output signal.

In this work, an open-loop clock generator for core frequency generation has been designed and implemented. It is based on a reverse phase switching scheme combined with programmable frequency dividers. It realizes sub-integer division ratios without fractional spurs in the output signal. It provides a wide range of division ratios with a low minimum ratio of 3, which is in contrast to previously published fractional-N PLL loop dividers which are designed for realization of consecutive division ratios. Furthermore this open-loop clock generator provides 50% output duty cycle which is essential for the targeted applicaiton within heterogeneous MPSoCs (e.g. for DDR2/3). It allows instantaneous changes of the division ratio within a single output clock cycle to realize fast core frequency changes for DVFS. Its functionality is explained in the following circuit description as presented in [HHES11].

5.1.2.1 Circuit Description

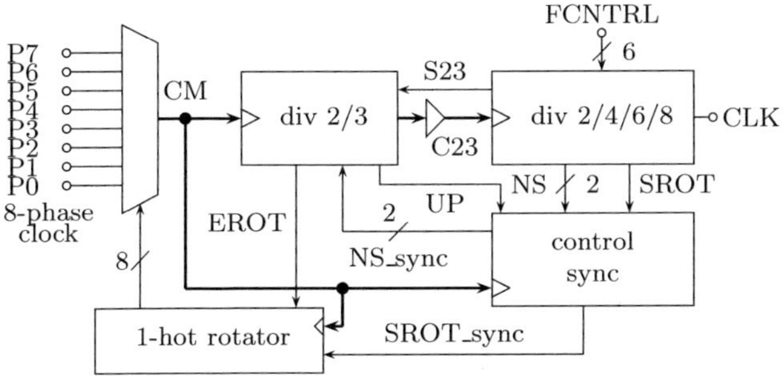

(a) block schematic

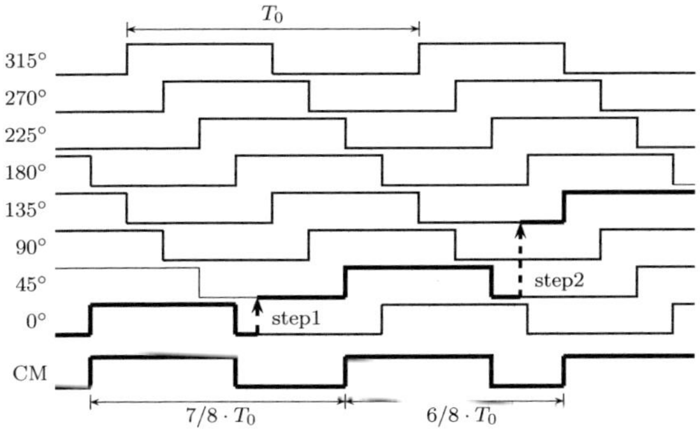

(b) reverse phase switching waveforms with 8-phase clock signal

Figure 5.5: Open-loop clock generator for frequency division based on reverse phase switching, [HHES11]

Toplevel Fig. 5.5(a) shows the toplevel schematic of the proposed open-loop clock generator. A phase multiplexer selects 1 out of $M = 8$ input clock phases with period T_0. This multiplexer output signal is divided by $N_{23} \in [2, 3]$ and by $N_{\text{sync}} \in [2, 4, 6, 8]$ in the output divider. So the base division ratio reads $N_{23} \cdot N_{\text{sync}}$ if no phase switching occurs.

The phases are switched in a glitch-free reverse switching scheme [SSS05, HSN09] as illustrated in Fig. 5.5(b). Each time the phase is switched the multiplexer output period is shortened. Following the theory in [Flo08] which suggests a switch step of $|n_{\text{step}}| \leq N/4 = 2$, we chose $n_{\text{step}} \in [1, 2]$ which leads to reduction of the multiplexer output period by $1/8 \cdot T_0$ or $2/8 \cdot T_0$ per switching event respectively. The multiplexer select signals are generated by a rotator that acts like an +1 or +2 adder compared to the flying adder frequency synthesis approach [XY03]. The multiplexer output clock CM is fed to the divide-by-2-or-3 circuit which generates the clock C23 and the enable pulse for the rotator. From $n_{\text{sw}} = 0$ up to $n_{\text{sw}} = 3$ phase switchings can occur per C23 cycle. The clock C23 is fed to the synchronous frequency divider with *even* division ratios N_{sync} that ensure 50% duty cycle of the output clock. In summary, the output core period reads

$$T_{\text{core}} = T_0 \cdot N_{\text{sync}} \cdot \left(N_{23} - \frac{n_{\text{sw}} \cdot n_{\text{step}}}{8} \right) \tag{5.1}$$

The open-loop clock generator is controlled by a 6-bit signal FCNTRL summarized in Tab. 5.1. For the 64 control words 33 different division ratios in the range from 3 to 24 can be realized as shown in Fig. 5.6. Only the integer primes 13, 17 and 19 are missing. For $T_0 = 500$ps the available output frequencies include 100MHz, 133MHz, 166MHz, 200MHz, 266MHz, 333MHz, 400MHz, 533MHz and 666MHz with 50% duty cycle for DDR, DDR2 and DDR3 memory interfaces [JED09, JED10]. In the following calculations we assume $T_0 = 500$ps as nominal operation frequency.

Table 5.1: Open-loop clock generator control signal definition

FCNTRL	comment	values
5:4	output division ratio	00: $N_{\text{sync}} = 0$; 01: $N_{\text{sync}} = 1$; 10: $N_{\text{sync}} = 2$; 11: $N_{\text{sync}} = 3$
3	divide by 2/3 select signal ($\overline{S23}$)	0: $N_{23} = 2$; 1: $N_{23} = 3$
2:1	switchings per C23 cycle (NS[1:0])	00: $n_{\text{sw}} = 3$; 01: $n_{\text{sw}} = 2$; 10: $n_{\text{sw}} = 1$; 11: $n_{\text{sw}} = 0$
0	rotate step (SROT)	0: $n_{\text{step}} = 2$; 1: $n_{\text{step}} = 1$

Phase multiplexer The most critical circuit component of the proposed clock generator is the phase multiplexer which selects 1 out of 8 input phases. The delay through the multiplexer $t_{\text{d,CIN}} \rightarrow \text{CM}$ must be as short as possible because the next select signal generated by the rotator (clocked with CM) must be settled before the

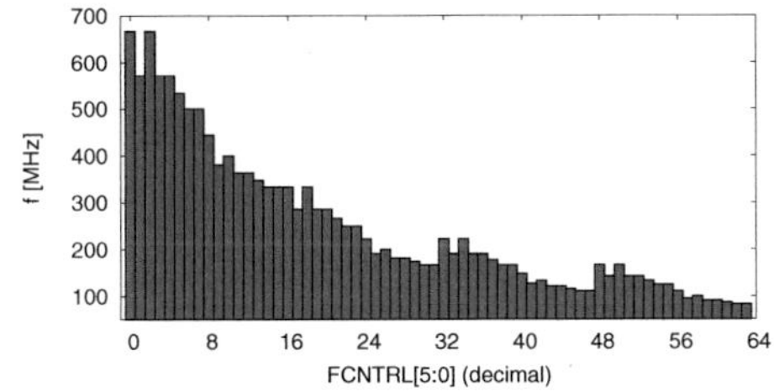

Figure 5.6: Available clock generator output frequencies for $T_0 = 500$ps, [HHES11]

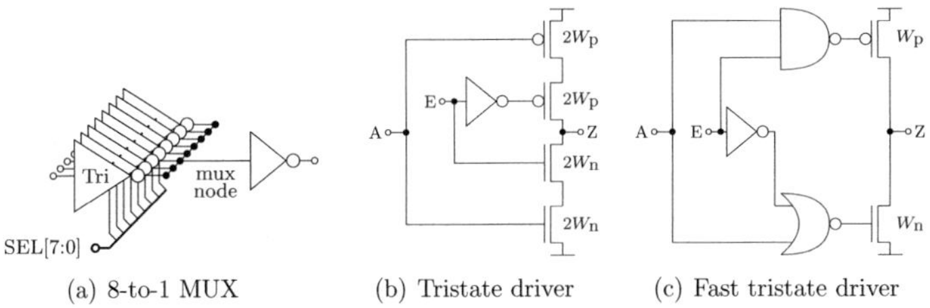

Figure 5.7: Phase multiplexer schematic

next rising clock edge arrives at the multiplexer input. For a $2GHz$ input signal and phase switching stepsize of 2 the constraint reads $t_{\mathrm{d,CIN}} \rightarrow \mathrm{S} < 375$ps. So only a single stage topology is feasible. Here a 8-to-1 tristate multiplexer as shown in Fig. 5.7(a) is used where 8 tristate drivers work on one multiplexer node. The total capacitance of the multiplexer node is given by

$$C_{\mathrm{muxnode}} = 8 \cdot C_{\mathrm{tri,out}} + C_{\mathrm{wire}} + C_{\mathrm{buf,in}}. \tag{5.2}$$

where $C_{\mathrm{tri,out}}$ is the output capacitance of a single tristate driver, C_{wire} is the wire routing capacitance and $C_{\mathrm{buf,in}}$ is the input capacitance of the output driver. In this design it is $C_{\mathrm{wire}} \approx 7.0$fF and $C_{\mathrm{buf,in}} \approx 2.4$fF. Fig. 5.7(b) shows the schematic of a conventional tristate driver, where two PMOS and NMOS devices are connected in series in the pull-up and pull-down path respectively. To achieve a drivestrenght equivalent to the PMOS width W_{P} and the NMOS width W_{N}, the actual device widths must be doubled ($2W_{\mathrm{P}}$, $2W_{\mathrm{N}}$) which leads to increased output capacitance (here $C_{\mathrm{tri,out,TRIX2}} \approx 2.3$fF). Therefore a high output drivestrength topology as shown in Fig. 5.7(c) is employed, where the gate signals of the driving PMOS and NMOS devices are generated by logic cells separately. This enables to achieve a similar output drivestrength as in Fig. 5.7(b) with reduced output capacitance (here

$C_{\text{tri,out,TRIFAST}} = 1.5\text{fF}$). Thus the $R_{\text{on,tri}} \cdot C_{\text{muxnode}}$ time constant can be reduced by 25%. Circuit simulations show that especially in the worst-case corner (SS process, $V_{\text{DD}} = 1.08V$, $T = 85°$) the delay through the multiplexer is dominated by $R_{\text{on,tri}} \cdot C_{\text{muxnode}}$, where the use of the high output drivestrength driver topology enables circuit operation. Fig. 5.8 shows the simulated waveforms of the multiplexer node for driver topology Fig. 5.7(b) and driver topology Fig. 5.7(c) respectively. The conventional tristate driver architecture fails under worst-case conditions whereas the use of the fast tristate driver increases the delay margin by 60ps and enables safe operation. The proposed topology slightly increases the area of the multiplexer circuit but does not increase its power consumption. The simulated power consumptions under worst-case timing conditions are 289.4uW for the fast multiplexer and 288.9uW for the conventional tri-state realization.

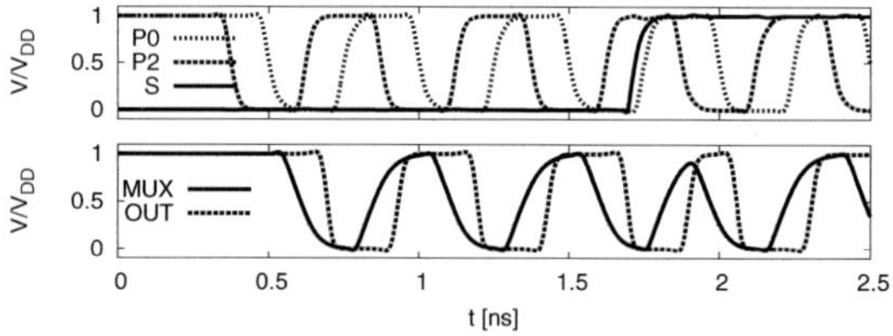

(a) Proposed tristate driver, switching from P4 to P6

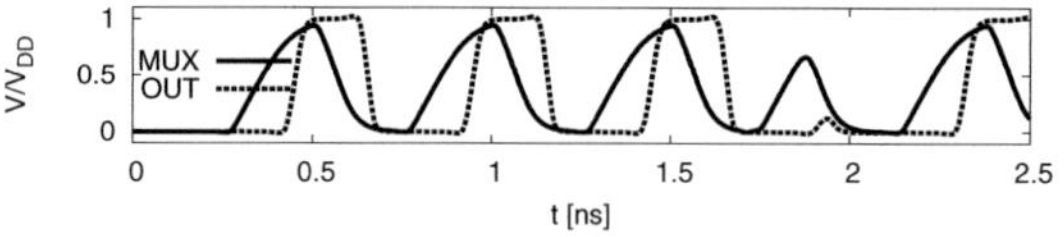

(b) Conventional tristate inverter, switching from P4 to P6

Figure 5.8: Simulated multiplexer phase switching waveforms in worst-case timing corner (SS 1.08V 85°), post layout, 65nm CMOS, [HHES11]

Divide-by-23 and rotator enable Fig. 5.9 shows the schematic of the divide by 2/3 circuit which is realized as a 2-bit state machine, driven with the clock CM. The state transfer equations are

$$S0‘ = S1 \tag{5.3}$$

$$S1‘ = \overline{(S23 + S0) \cdot S1} \tag{5.4}$$

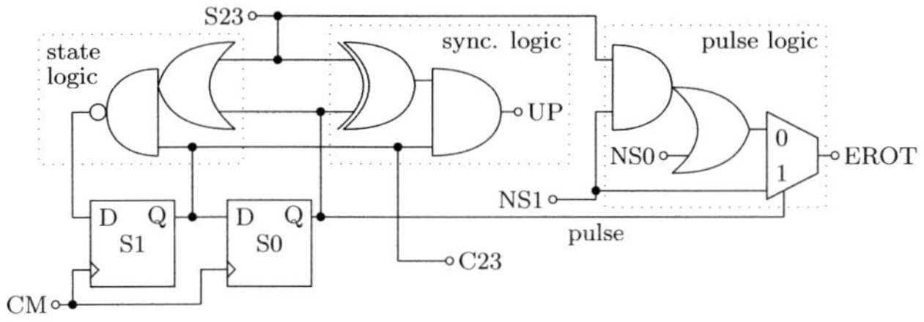

Figure 5.9: Divider by 2 and 3 and rotate pulse generation logic, [HHES11]

leading to a divide-by-2 state sequence of 10 → 01 → 10... if $S23 = 1$ and a divide-by-3 state sequence of 10 → 11 → 01 → 10... if $S23 = 0$. The output clock C23 is directly derived from S1 (C23 = S1). This circuit generates enable pulses EROT for the phase rotator depending on n_{sw} (NS1 and NS2) as shown in Tab. 5.1. During one C23 cycle 0 to 3 enable pulses can be generated. EROT is defined by

$$\mathrm{EROT} = \mathrm{S0} \cdot \mathrm{NS1} + \overline{\mathrm{S0}} \cdot (\mathrm{NS0} + \mathrm{NS1} + \mathrm{S23}). \tag{5.5}$$

Additionally, this state machine synchronizes the control signals for the rotator step size and the number of phase switchings by generation of an update signal UP. The rising edge of C23 occurs two cycles after UP is set to 1, with respect to the selected division ratio S23

$$\mathrm{UP} = \mathrm{S23} \cdot \overline{\mathrm{S0}} \cdot \mathrm{S1} + \overline{\mathrm{S23}} \cdot \mathrm{S0} \cdot \mathrm{S1} \tag{5.6}$$

which ensures in combination with the control synchronizer shown in Fig. 5.10(a) that SROT_sync and NS_sync are constant within each C23 cycle. Fig. 5.10(b) shows an example state sequence where the division ratio is changed from 2 to 3 (on rising edge C23) and the number of switchings from 0 to 3 (on rising edge CM, enabled by UP). The critical timing constraint of this circuit is the setup time of S23, which is generated in the synchronous output divider with rising edge of C23, that must be settled before the next rising edge of CM.

Rotator Fig. 5.11 shows the schematic of the 1-hot rotator that generates the phase multiplexer select signals. It consists of a closed-loop shift register, of which one flip-flop has a reset state 1 and all others 0. This approach ensures synchronous select signals with minimum skew and avoids glitches at the phase multiplexer. The enable signal EROT activates a clock gate. If no phase rotation occurs the shift register is not clocked which reduces power consumption. Multiplexers are used to

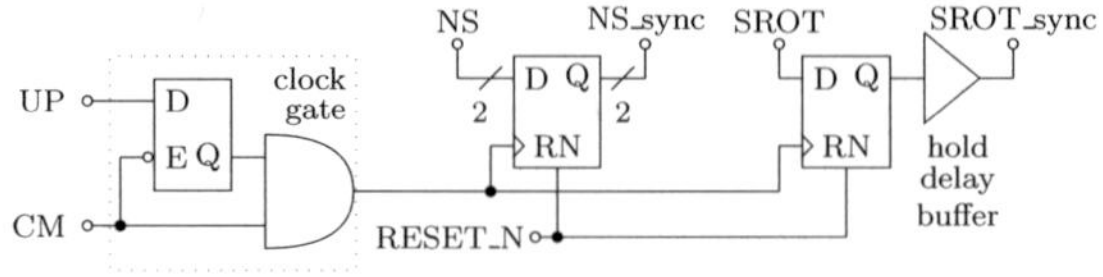

(a) Control signal synchronizer

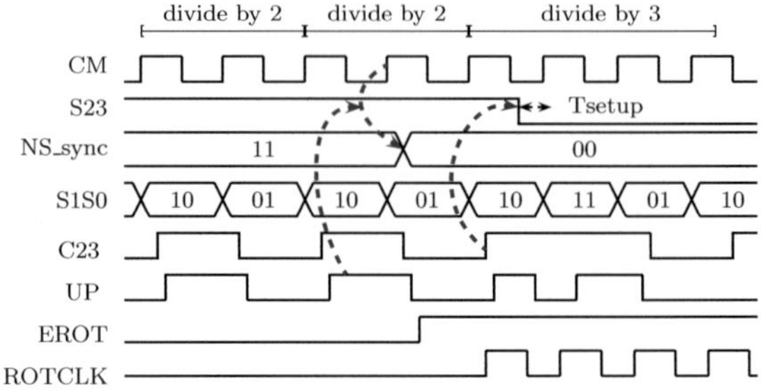

(b) State sequence of Div 23, Change of S23 and NS1 and NS2

Figure 5.10: Synchronization of frequency division control signals, [HHES11]

control the rotation step which can be 1 or 2 depending on the control signal SROT. This realization enables minimum combinational logic between the flip-flops which is mandatory concerning its worst-case CM clock period of 375ps.

Synchronous output divider Fig. 5.12(a) shows the schematic of the synchronous frequency divider. It provides division ratios of 2, 4, 6 and 8 with 50% output duty cycle using a 3-bit state machine running at clock C23. The state sequences are

div2: $0 \rightarrow 4 \rightarrow 0 \ldots$

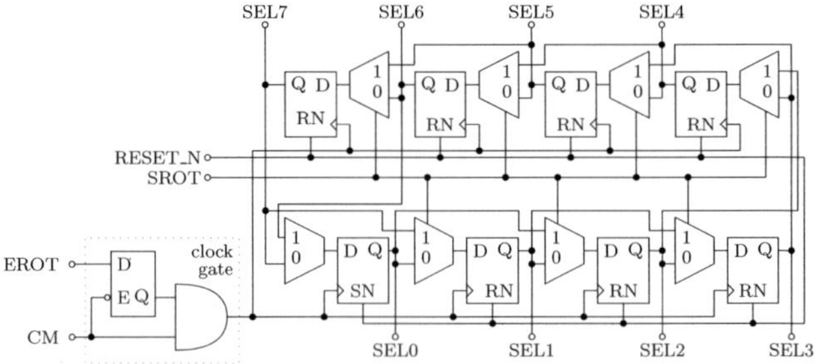

Figure 5.11: 1-hot rotator schematic, [HHES11]

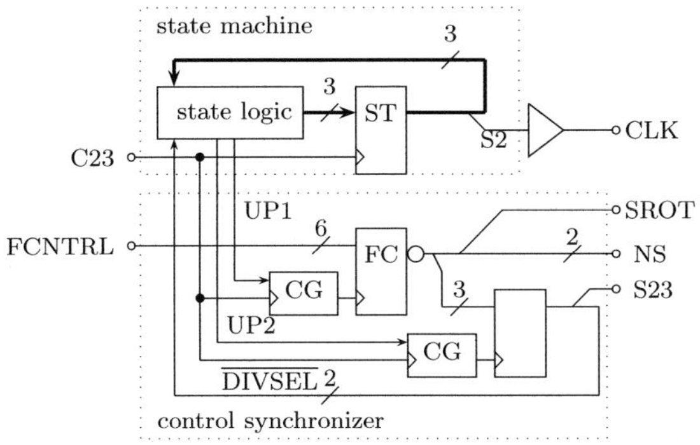

(a) schematic

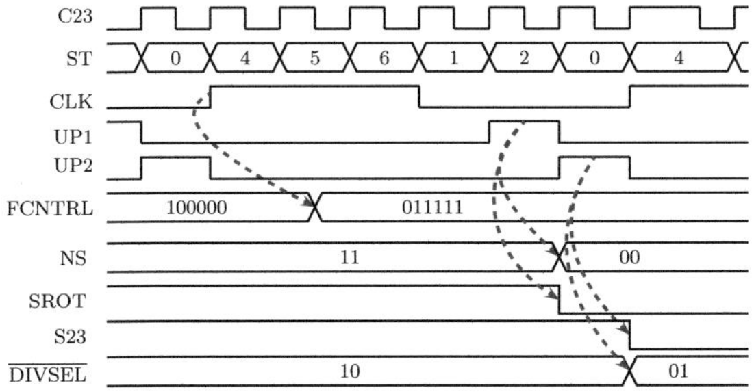

(b) state sequence example (div6)

Figure 5.12: Synchronous output divider and control synchronizer, [HHES11]

div4: $0 \rightarrow 4 \rightarrow 6 \rightarrow 2 \rightarrow 0 \ldots$

div6: $0 \rightarrow 4 \rightarrow 5 \rightarrow 6 \rightarrow 1 \rightarrow 2 \rightarrow 0 \ldots$

div8: $0 \rightarrow 4 \rightarrow 5 \rightarrow 6 \rightarrow 7 \rightarrow 1 \rightarrow 2 \rightarrow 3 \rightarrow 0 \ldots$

The output clock CLK is directly derived from the state MSB. Its rising edge occurs at the state transition $0 \rightarrow 4$ independent from the selected division ratio DIVSEL. Additionally this circuit synchronizes the internal control signals. The frequency control input FCNTRL is captured in register FC at state 0 enabled by the update signal UP1. The S23 control signal is delayed by 1 C23 cycle to be updated with the rising edge of the output clock (state 4) enabled by UP2. This ensures together with the control signal synchronizer shown in Fig. 5.10(a) that the internal control signals remain constant within one output clock cycle of the open-loop clock generator. Fig. 5.12(b) shows an example state sequence of the synchronous output divider with the corresponding control signal updates. When the frequency control input FCNTRL is updated with the rising edge of CLK (state 4) by an external register, it must be settled before state 0 with a certain setup margin of the sampling register FC.

This constraint is modeled in the Liberty (.lib) file of the open-loop clock generator macro. Based on the functional circuit timing (Fig. 5.13(a)) the black box .lib timing model is defined as shown in Fig. 5.13(b) and Fig. 5.13(c). An internal virtual clock signal (clk_int) is generated as clock root pin. From this the virtual FC flip-flop clock is generated with a delay $t_{\mathrm{D,CG}}$, modeling the clock gate delay, and the output clock CLK is generated with the output delay $t_{\mathrm{D,O}}$. The worst-case timing conditions with respect to the frequency control input FCNTRL are:

- The synchronous output divider has a division ratio of 2, where the FC capture clock (in state 0) corresponds to the falling edge of the output clock.
- The output clock has its shortest period $T_{\mathrm{CLK}} = 1.5\mathrm{ns}$.

Therefore in the .lib file model the FCNTRL inputs are constrained with setup (t_{S}) and hold (t_{H}) with respect to the *falling* edge of the internal pin clk_FC, where t_{S} and t_{H} are the setup and hold times of the standard cell flip-flops in the FC register. Details on application of these constraints for integration into the semi-custom digital design synthesis and place & route flow are given in Sec. 5.2.

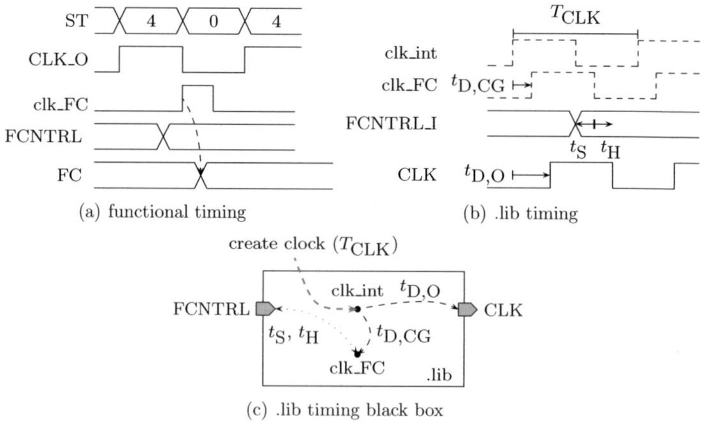

Figure 5.13: Timing model of the open-loop core clock generator

5.1.2.2 Implementation Results

The open-loop clock generator has been implemented in both 65nm and 28nm CMOS technology with identical logic circuit structure. The layouts of the implementations are shown in Fig. 5.14. A high-speed standard cell library is used for circuit implementation. Special customized cells (e.g. fast tri-state driver) are used for timing critical paths. To achieve good phase symmetry and reduce mismatch induced jitter at the output, the rotator is merged with the phase multiplexer. Thereby each rotator flip-flop is located next to the corresponding tristate driver to minimize delay on the select signal line and to ensure symmetry of the 8 clock phase inputs.

Fig. 5.15 shows the simulated power consumption of the clock generator for different PVT corners, and measurement results. The power consumption increases with the number of phase switchings due to the phase rotator activity. For some frequencies as shown in Fig. 5.6 different control signal realizations exist with different power consumption. This must be considered when selection the desired output frequency in the system application. This fully-static CMOS implementation of the open-loop clock generator scales well with technology in terms of area and power consumption.

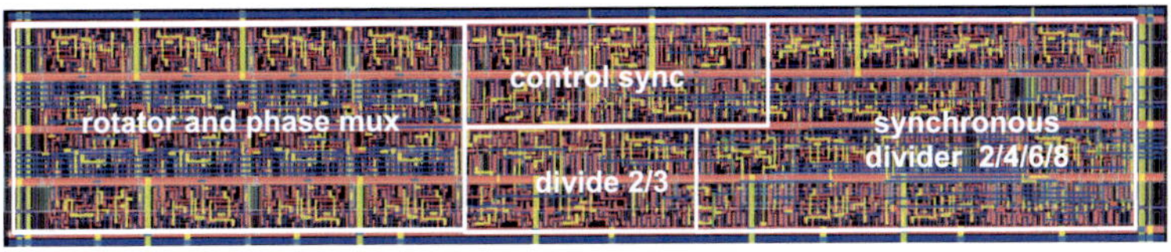

(a) 65nm, 63.2μm $\times$ 11.8μm

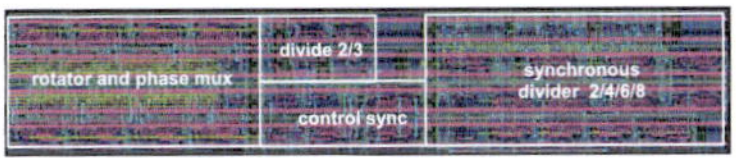

(b) 28nm, 29.8μm $\times$ 5.6μm

Figure 5.14: Open-loop clock generator layouts

Fig. 5.16 shows the measured output waveforms of the open-loop clock generator realization in 65nm CMOS technology on the "Tommy" testchip. Its input clock is provided by the ADPLL presented in Sec. 4.2. Fig. 5.17(a) shows the measured rms period jitter σ_T and the relative jitter $\sigma_T/T_{\mathrm{CLK}}$ and Fig. 5.17(b) shows the measured duty cycle which is slightly degraded by the clock measurement LVDS pad at higher output frequencies. The core clock jitter of the 28nm implementation is shown in Fig. 5.18. Here some frequency settings show increased jitter. This effect is analyzed in Sec. 5.1.3. Fig. 5.19 shows the output clock signal of the open-loop clock generator

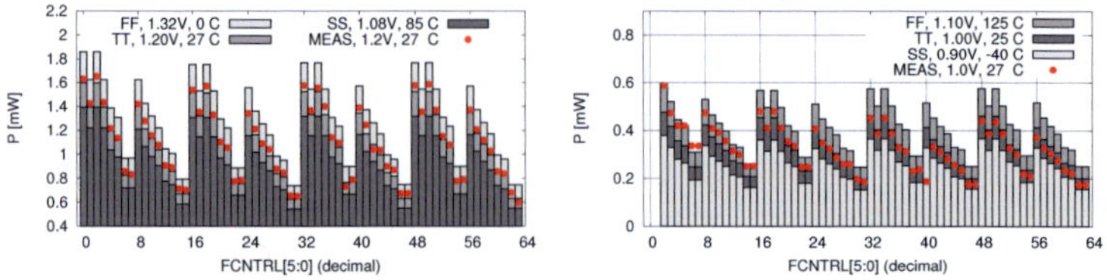

(a) 65nm (b) 28nm, (FCNTRL 0 and 1 not used)

Figure 5.15: Simulated and measured power consumption of the open-loop clock generator, $T_0 = 500\text{ps}$

when changing the frequency control word. This demonstrates the capability of arbitrary, instantaneous frequency changes. Fig. 5.20 shows instantaneous output frequency changes of the 28nm implementation on the "Cool28SoC" testchip.

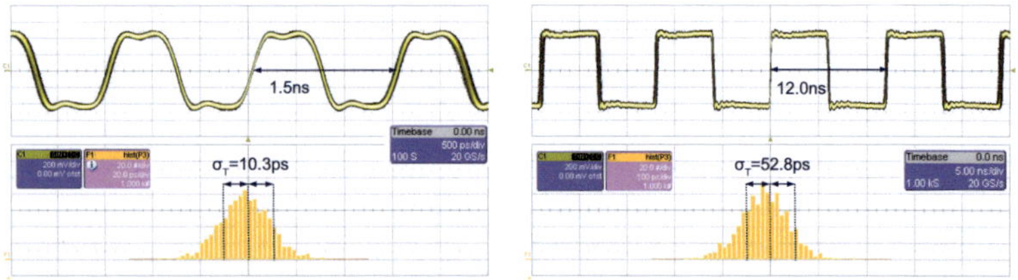

(a) 666MHz (b) 83MHz

Figure 5.16: Measured output signals and period jitter histograms of the 65nm open-loop clock generator realization at maximum and minimum output frequency, [HHES11]

As presented in [HHES11], Tab. 5.2 summarizes the performances of recently published frequency dividers and open-loop clock generators in CMOS technologies $\leq$ 180nm for comparison with this work. The Flying Adder synthesizer reported in [Xiu07] provides many different output frequencies but the maximum input frequency and therefore the maximum output frequency is limited due to the arithmetic logic required. The sequential divider in [BSS$^+$08] has a high maximum input frequency but also high, only integer, division ratios. [WYZF07] presents a sequential divider with wide division ratio range but only integer division ratios. [CC08] uses a wide phase multiplexer (32 inputs) and a modulus counter to achieve sub-integer division ratios but allows only small maximum input frequencies.

The proposed design enables high maximum input frequency, as usually achieved in sequential or phase rotating dividers, combined with a small minimum division ratio as achieved by the Flying Adder approach. It shows low power consumption

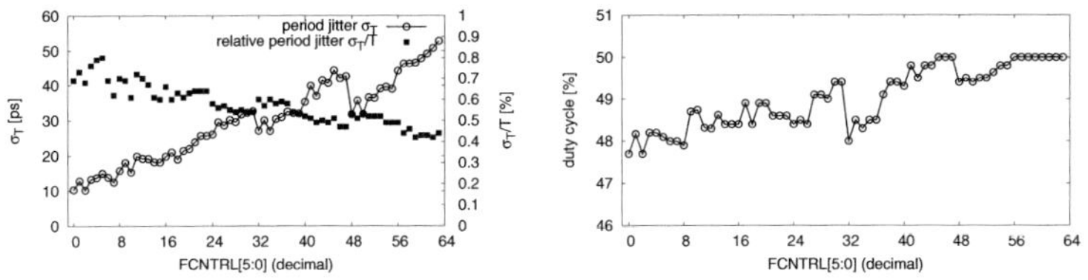

(a) jitter measurement results (b) duty cycle

Figure 5.17: Clock quality measurement results, 65nm, "Tommy" testchip

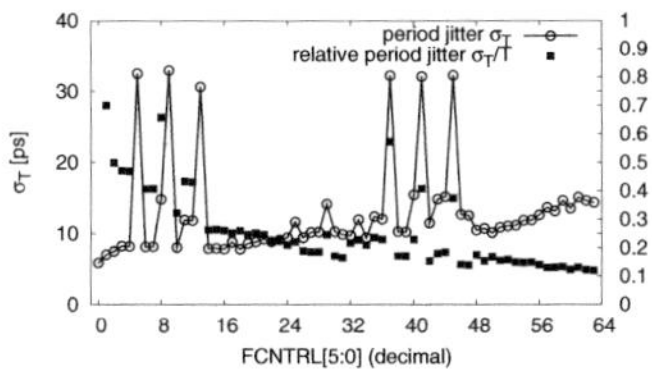

Figure 5.18: Core clock jitter measurement result, 28nm, "Cool28SoC" testchip

and requires only small chip area. By its purely static CMOS logic implementation it scales well with the shrinking of semiconductor technologies.

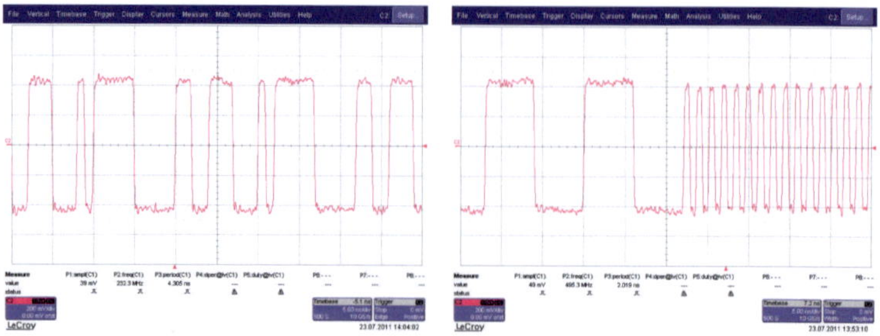

(a) Rotation between 4 different output period settings of 2ns, 10ns, 4ns and 6ns

(b) Change from 12ns to 1.5ns output period

Figure 5.19: Measured instantaneous output period changes of the open-loop clock generator, 65nm realization, "Atlas" testchip

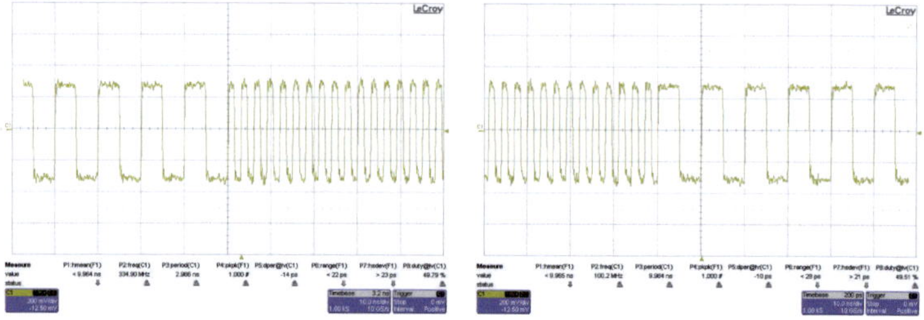

(a) Change from 10ns to 3ns output period

(b) Change from 3ns to 10ns output period

Figure 5.20: Measured instantaneous output period changes of the open-loop clock generator, 28nm realization, "Cool28SoC" testchip

Table 5.2: Open loop clock generator design comparison

ref	**type** [5]	**tech.** [nm]	$\mathbf{f_{in}}$ [MHz]	$\mathbf{f_{out,min}}$ [MHz]	$\mathbf{f_{out,max}}$ [MHz]	$\mathbf{N_{min}}$	$\mathbf{N_{max}}$	**# N**	**area** [μm^2]	**power** [mW]
[Xiu07] [6]	fa	90	1269	100	250	n.a.	n.a.	60	151000 [7]	n.a.
[BSS+08]	seq	90	3500	130	146	24	27	4	21800	4.5
[WYZF07]	seq	180	1500	5.8	750	2	256	255	n.a.	1.3
[CC08]	pr	180	238	7.4	122.6	2	32	frac	91000	48
this work	pr	65	2000	83	667	3	24	33	744	0.62 to 1.6
this work	pr	28	2000	83	667	3	24	33	167	0.20 to 0.47

[5] fa: flying adder, pr: phase rotator, seq: sequential divider

[6] ARM/DDR frequency synthesizer

[7] with PLL

5.1.3 Period Jitter Analysis

The output signal of open-loop clock generation circuits exhibits period jitter which results from different sources. First, the noise at the ADPLL DCO output is accumulated during frequency division. The output period of the open-loop clock generator is the sum of N DCO cycles, N is the (sub-integer) division ratio of the open-loop clock generator. Assuming DCO noise from white thermal sources, the standard deviation of the divider output clock period can be estimated by

$$\sigma_{T,\mathrm{core}} = \sqrt{N} \cdot \sigma_{T,\mathrm{DCO}}. \tag{5.7}$$

where $\sigma_{T,\mathrm{DCO}}$ is the standard deviation of the DCO period jitter. This also holds for $N < 1$ in case of the open-loop frequency multiplication [vdBKVN02], where the DCO control signal influence on the period jitter is neglected. This assumption holds for the targeted application with an ADPLL as presented in Sec. 4.2 where the DCO tuning signal is updated only with the lower frequency reference clock. Second, the logic of the open-loop clock generator adds jitter due to device noise within its logic cells. However, this jitter does not accumulate over multiple clock cycles.

Besides the noise induced jitter, the open-loop clock generators are sensitive to static mismatch within their multi-phase input signals. This can be caused by systematic mismatch due to non-symmetric layout realizations or local on-chip variations in the DCO. An analysis of the phase-mismatch related jitter of the open-loop clock generator is presented in [HEH+13] and explained in the following.

The multi-phase input clock is generated by the DCO with a chain of delay cells which are locked to a period of T_0 by the closed loop ADPLL. The DCO with four differential stages provides $M = 8$ phases, where the delays are represented by the rising edge delays and falling edge delays of each delay stage. Each stage exhibits a delay which can be written as

$$\Delta t = c \cdot t_{\mathrm{d},0} \cdot (1 + \epsilon_{\mathrm{step}}) \tag{5.8}$$

where ϵ_{step} is the relative variation which is assumed to be distributed normally with standard deviation σ_{step} and zero mean value. c represents a normalized tuning constant which allows tuning of the delay cell around its nominal delay value of $t_{\mathrm{d},0}$

by the ADPLL. For M phases the total delay is locked to the reference T_0

$$\sum_{i=1}^{M}(c \cdot t_{\mathrm{d},0} \cdot (1+\epsilon_{\mathrm{step},i})) = T_0. \tag{5.9}$$

So the resulting tune control value c reads

$$c = \frac{T_0}{t_{\mathrm{d},0} \cdot \left(M + \sum_{i=1}^{M} \epsilon_{\mathrm{step},i}\right)}. \tag{5.10}$$

Note that each physical delay cell is considered here with two delays, representing the rising and falling transition within one clock period respectively. Their variation is modeled independently. The time difference of a switching step over n phases is

$$t_{\mathrm{step},n} = c \cdot t_{\mathrm{d},0} \cdot \left(n + \sum_{i=1}^{n} \epsilon_{\mathrm{step},i}\right) \tag{5.11}$$

$$= \frac{T_0}{M} \cdot \frac{\left(n + \sum_{i=1}^{n} \epsilon_{\mathrm{step},i}\right)}{\left(1 + \sum_{i=1}^{M} \frac{\epsilon_{\mathrm{step},i}}{M}\right)} \tag{5.12}$$

by using Eq. 5.10. Assuming that $\epsilon_{\mathrm{step},i} \ll 1$ and $\sigma_\epsilon \ll 1$ Eq. 5.12 can be approximated by

$$t_{\mathrm{step},n} \approx \frac{T_0}{M} \cdot \left(n + \sum_{i=1}^{n} \epsilon_{\mathrm{step},i}\right) \cdot \left(1 - \sum_{i=1}^{M} \frac{\epsilon_{\mathrm{step},i}}{M}\right) \tag{5.13}$$

$$\approx \frac{T_0}{M} \cdot \left(n + \sum_{i=1}^{n} \epsilon_{\mathrm{step},i} - n \cdot \sum_{i=1}^{M} \frac{\epsilon_{\mathrm{step},i}}{M}\right) \tag{5.14}$$

$$= \frac{T_0}{M} \cdot \left[n - \left(\sum_{i=1}^{n}\left(\frac{n}{M} - 1\right) \cdot \epsilon_{\mathrm{step},i} + \sum_{i=n+1}^{M} \frac{n}{M} \cdot \epsilon_{\mathrm{step},i}\right)\right]. \tag{5.15}$$

Since $T_0 \cdot \frac{n}{M}$ denotes the nominal value of $t_{\mathrm{step},n}$, its variation can be expressed by

$$\Delta t_{\mathrm{step},n} \approx \frac{T_0}{M} \cdot \left(\sum_{i=1}^{n}\left(\frac{n}{M} - 1\right) \cdot \epsilon_{\mathrm{step},i} + \sum_{i=n+1}^{M} \frac{n}{M} \cdot \epsilon_{\mathrm{step},i}\right). \tag{5.16}$$

Considering uncorrelated variations of the delay steps, its standard deviation reads

$$\sigma_{t,\mathrm{step},n}^2 \approx \frac{T_0^2}{M^2} \cdot \left(n \cdot \left(\frac{n}{M} - 1\right)^2 \cdot \sigma_\epsilon^2 + (M-n)\frac{n^2}{M^2} \cdot \sigma_\epsilon^2\right) \tag{5.17}$$

which can be simplified to

$$\sigma^2_{t,\mathrm{step},n} \approx \frac{T_0^2}{M^2} \cdot \sigma_\epsilon^2 \cdot \left(n - \frac{n^2}{M}\right). \tag{5.18}$$

The standard deviation of a single delay element is

$$\sigma_{t,\mathrm{step}} = \frac{T_0^2}{M^2} \cdot \sigma_\epsilon^2. \tag{5.19}$$

Thereby the sensitivity of the total phase switching timing error over n steps with respect to the timing variation of a single delay element reads

$$S_{\mathrm{core}} = \frac{\sigma_{T,\mathrm{core}}}{\sigma_{t,\mathrm{step}}} = \sqrt{n - \frac{n^2}{M}}. \tag{5.20}$$

In case of the open-loop clock generator for frequency multiplication presented in Sec. 5.1.1, the high-speed clock is generated by combining two differential clock phases with spacing of $4t_{\mathrm{step}}$. So the mismatch jitter sensitivity of the doubled DCO clock reads

$$S_{\mathrm{NoC}} = \frac{\sigma_T}{\sigma_{t,\mathrm{step}}} = \sqrt{4} = 2. \tag{5.21}$$

For the open-loop core clock generator the switch timing error for n effective switchings per output cycle directly translates to period jitter of its output signal

$$\sigma_{T,\mathrm{core}} = \Delta t_{\mathrm{step},n}. \tag{5.22}$$

Fig. 5.21 shows the mismatch sensitivity for different M and n. The total number of

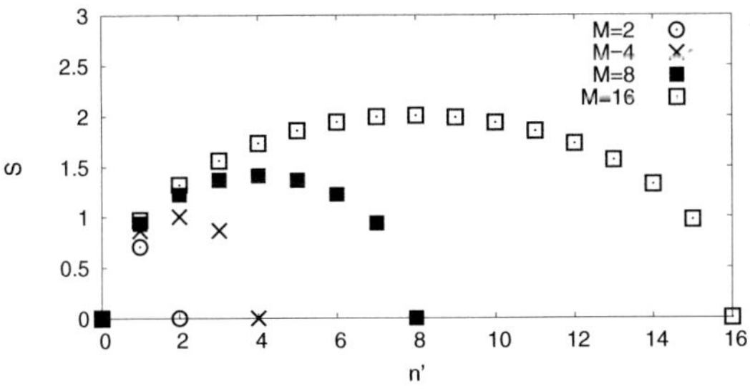

Figure 5.21: Timing error sensitivity of multi-phase oscillator outputs

switchings within one open-loop clock generator output period for a given division ratio as presented by Eq. 5.1 is $n_{\text{switchings}} = N_{\text{sync}} \cdot n_{\text{sw}} \cdot n_{\text{step}}$ resulting in an effective number of switchings

$$n = \text{mod}_8 \left(N_{\text{sync}} \cdot n_{\text{sw}} \cdot n_{\text{step}} \right). \tag{5.23}$$

The measured the output jitter of the open-loop clock generator includes both, components from noise sources and components due to static phase mismatch. From the nominal open-loop frequency output period $T = N \cdot T_0$ and the DCO jitter accumulation from DCO noise shown in Eq. 5.7 it can be concluded, that $\sigma_{T,\text{core}}/\sqrt{T_{\text{core}}}$ should be constant, if only DCO jitter is accumulated, if the open-loop clock generator is noise free and if no phase mismatch is present. In the real circuit phase mismatch leads to increased output jitter if $n_{\text{eff}} \neq 0$ for high output frequencies (low FCNTRL). For lower output frequencies the phase mismatch effect decreases because if k DCO periods are summed, the output jitter is $\sigma_{T,\text{core}} \propto \sqrt{k}$, whereas the phase mismatch due to n_{eff} remains constant. The open-loop clock generator internal noise adds jitter as well which results in increased $\sigma_{T,\text{core}}/\sqrt{T_{\text{core}}}$ for lower frequencies.

To illustrate this effect Fig. 5.22 shows the measured normalized period jitter $\sigma_{T,\text{core}}/\sqrt{T_{\text{core}}}$ of the core clock together with the number of effective phase switchings n_{eff} from Eq. 5.23 and the mismatch jitter sensitivity S from Eg. 5.20. The measurements have been performed on multiple chips. It can be seen that the phase mismatch effect is present, especially at higher output frequencies, but does not significantly degrade the output clock quality.

The jitter caused by device noise within the circuit components of the open-loop clock generator has been simulated using transient noise analyses for both the 65nm and 28nm circuit implementations. In both cases the additional period jitter rms value is below 0.6ps and therefore significantly lower than the phase mismatch effect.

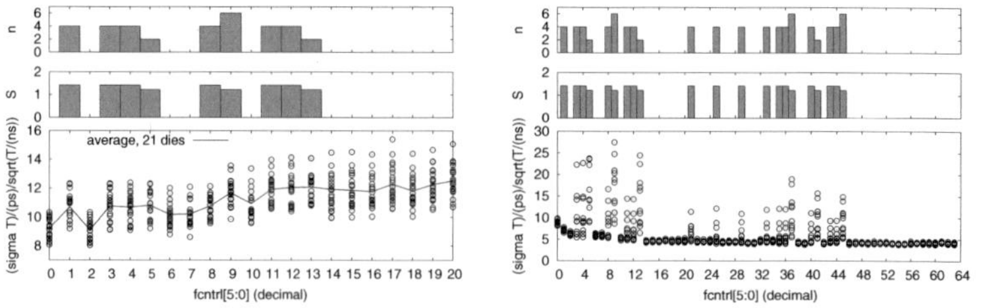

(a) 65nm CMOS ("Tommy"), 21 dies measured (b) 28nm CMOS ("Cool28SoC"), 14 dies measured

Figure 5.22: Measured mismatch jitter of the open-loop clock generator

The measurement results of the 28nm realization show significantly increased jitter, at those specific frequency control settings which are sensitive to mismatch in the multi-phase input signals. However, the noise influence of the open-loop clock generator is negligible, which can be seen by the flat normalized jitter in Fig. 5.22(b) at those frequency settings which are not sensitive to mismatch.

Deeper analysis of the increased mismatch jitter revealed, that it is caused by the level shifting CMOS inverter within the 28nm DCO as presented in Sec. 3.2, between the tuning voltage $V_{\mathrm{DD,tune}}$ and the nominal supply. Fig. 5.23 shows the waveform of one Monte-Carlo simulation sample, showing the multi-phase output of the 28nm DCO, which is fed to the open-loop clock generator. The falling signal edges (0-to-1 transition in the $V_{\mathrm{DD,tune}}$ domain) are significantly slowed down and thereby heavily effected by mismatch of the output inverters. In contrast, the rising edges (1-to-0 transition in the $V_{\mathrm{DD,tune}}$ domain) show steeper slopes and less mismatch. Fig. 5.24(a) shows the resulting mismatch jitter histogram of the open-loop clock generator output at settings FCNTRL=5, which is in good agreement with the measurement results in Fig. 5.22(b).

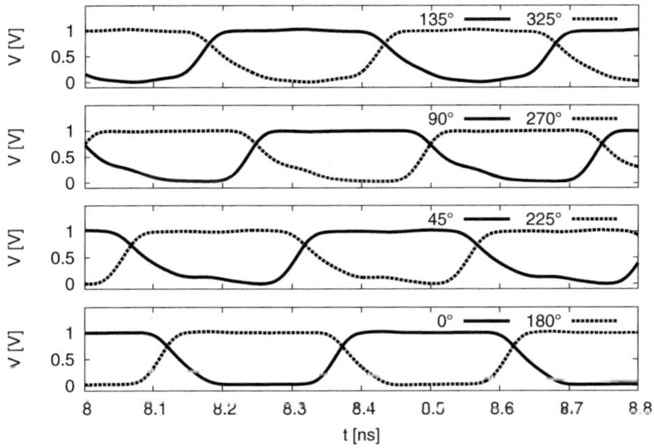

Figure 5.23: Simulated waveform of 28nm DCO multi-phase output with mismatch, 1.0V, 25C, global and local variations

This phenomenon and a possible solution to this issue by using cross-coupled clock buffers has been presented in Sec. 3.4, which align the rising and falling clock edges based on the 180° shifted phases. However, this is not required here since the open-loop clock generator architecture is only sensitive to one single clock edge. All internal switching of sequential elements is triggered by the rising clock edge, as

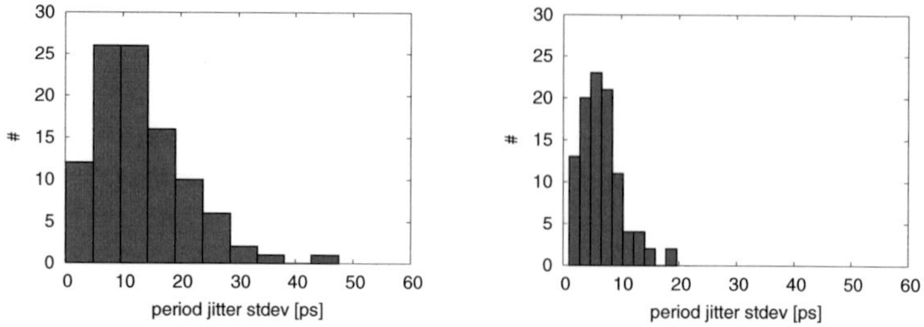

(a) initial version, falling edge sensitive (b) improved version, rising edge sensitive

Figure 5.24: Simulated 28nm open-loop clock generator, output period jitter sigma at FCNTRL=5, 1.0V, 25C, Monte-Carlo simulation with global and local variations

shown in Fig. 5.1, which is the falling edge of the multi-phase input clocks due to an additional input inverter. Therefore the addition of cross coupled clock buffers can be prevented, which does not increase power consumption.

The modification which is applied here to circumvent this issue, is to invert the edge-sensitivity of the open-loop clock generator compared to the 65nm realization. This is done by adding an inverter at the phase multiplexer output and re-connecting the multi-phase inputs by 180 degree phase shift, as shown in Fig. 5.25. This significantly improves the mismatch related output jitter, as shown in the Monte-Carlo simulation result in Fig. 5.24(b), as example.

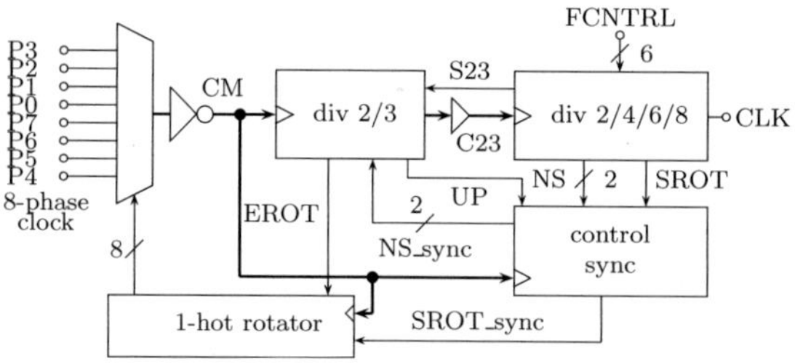

Figure 5.25: Open-loop clock generator schematic, improved version in 28nm design, rising edge sensitive

5.2 MPSoC System Integration

5.2.1 Clock Generator Wrapper

Fig. 5.26 shows a block level schematic of the system integration concept of the proposed clock generators to the GALS MPSoC core wrapper as presented in Sec. 2.5. A clock generator wrapper encapsulates all clock generation circuits. It contains the ADPLL circuit and logic for core frequency selection based on the PL defined by the PMU. This is realized by a lookup table (LUT), which translates the PL code (e.g. 2-bit for 4 PLs) into the clock generator specific control signal (e.g. 6-bit FCNTRL for the open-loop clock generator as presented in Sec. 5.1.2). Additionally, the ADPLL provides the high-speed clock for serial NoC interface as presented in Sec. 2.4.

As explained in Sec. 2.7, ADPLL based clock generators are realized as macro blocks which are instantiated within the clock generator wrappers. This eases design implementation, because the ADPLL circuits are implemented once and can be used multiple times within the MPSoC. The clock generator wrapper itself is realized as parameterizable RTL description which is implemented by automated synthesis and place&route together with the GALS core wrapper. Parameters can be defined individually per core and for example include the number of PLs and the reset values of the PL lookup table.

Moreover, this design hierarchy allows for flexible architecture adoptions individually per core. As example, special handling of clocks and resets for design for test (DFT), can be completely realized in the clock generator wrapper, without specific changes within the ADPLL macro. The ADPLL macro itself is scannable using the reference clock signal in DFT mode. For special test architectures, like for example at-speed delay tests of the functional cores, the ADPLL clock generator is fully operational while the core logic is tested at the target application clock speed. The DCO within the ADPLL core is supplied by a dedicated supply voltage and ground net to reduce power supply noise in the sensitive DCO components for low jitter. The digital ADPLL components (e.g. controller, filter, frequency divider) are connected to the global digital power supply domain.

For automated synthesis and place&route implementation of the clock generator wrapper, timings have to be constrained. This includes both, the clock creation for the processor core and the interface timing of the clock generator itself for frequency changes. The timings of custom macro blocks (e.g. the open-loop clock generators) are modeled using Liberty (.lib) files, such that these blocks appear as black boxes in the timing analysis. The detailed interface timing of the open-loop clock generator

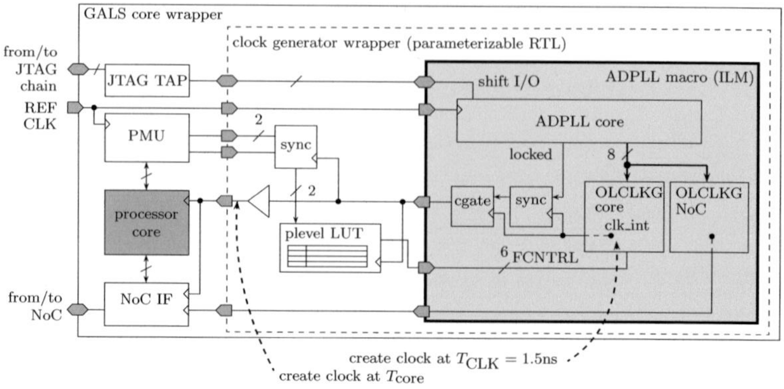

Figure 5.26: System integration schematic of the local clock generator for GALS MPSoCs

has been presented in Fig. 5.13 in Sec. 5.1.2.1 of this work. While this constraining for glitch free changes of the frequency requires the creation of the internal clock clk_int with the minimum period (e.g. $T_{CLK} = 1.5$ns), the processor core clock must be constrained separately with a period T_{core} based on the design target for the particular core content. T_{core} is usually larger than T_{CLK}. This issue is solved by the creation of two separated clocks for the internal clock generator operation and for the MPSoC core for timing analysis as shown in Fig. 5.26. As shown in Sec. 2.7, the ADPLL macro is implemented using an automated synthesis and place&route flow and is abstracted as ILM for integration into the clock generator wrapper.

This clock generator wrapper approach has been used in the testchips presented in Sec. 2.8. As example Fig. 5.27 shows a measurement results from "Atlas", where a DVFS PL change from (0.9V, 100MHz) to (1.2V, 333MHz) is performed within only 20ns. The supply voltage switching is controlled by the PMU as shown in Sec. 2.2.1 and the instantaneous frequency change is realized by the open-loop clock generator.

5.2.2 Clock Generator Integration Overhead

When exploring a suitable MPSoC architecture for a target application the power consumption of the local clock generators must be considered. It is desired that their power consumption is significantly lower than the core power to achieve real over all energy savings from this GALS architecture. In particular the power savings which are obtained from advanced fine-grained power management like DVFS should be higher than the overhead of local clock generation. However, this sig-

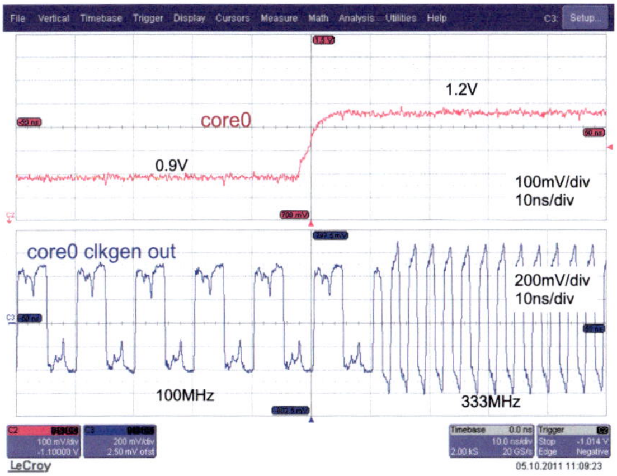

Figure 5.27: Measured supply voltage and clock waveform at PL change within 20ns, "Atlas" testchip

nificantly depends on the logic content of the cores and their target application scenario. Details can be evaluated during system design by means of silicon virtual prototyping [DHCC03] of the MPSoC including its power management architecture [AF10, AF11].

Table 5.3: MPSoC core power consumption examples

ref	tech	core (task)	f [MHz]	V_{DD} [V]	P [mW]	comment
[WKA+12]	65nm	FEC(LDPC)	267	1.20	367	measured on "Tommy"
[WKA+12]	65nm	FEC(Turbo)	333	1.20	283	measured on "Tommy"
[WKA+12]	65nm	Sphere Decoder	333	1.20	38	measured on "Tommy"
[WKA+12]	65nm	Sphere Decoder	381	1.30	50	measured on "Tommy"
	28nm	Tensilica® Xtensa LX4 DSP (Dhrystone benchmark)	500	1.00	30.1	measured on "Cool28SoC"

Tab. 5.3 summarizes some examples of the MPSoC core power consumptions of circuits which have been realized within the CoolBaseStations [EMF+12] and Cool-RF-28 projects. The relative power consumption of the 65nm ADPLL clock generators, which consumes 2.9mW from 1.2V, with respect to the power consumption of the hardware accelerators (FEC, Sphere Decoder) is suitably low. The power consumption of the 28nm ADPLL implementation of 0.64mW is low compared to

the power consumption of the DSP core on "Cool28SoC".
However, the clock generator power overhead can be further reduced when taking into account the modular architecture of the closed-loop ADPLL in combination with the open-loop clock generator. It is possible to share the closed-loop ADPLL with multi-phase clock outputs for several open-loop clock generators used for different cores or interfaces. Besides the standard application for a single core and NoC link as shown in Fig. 5.26, some additional application scenarios of the proposed clock generators are summarized in Fig. 5.28. Also switching off ADPLL of cores which are in idle state can significantly reduce the power consumption overhead of per-core instantiated GALS clock generators. In this scenario the fast lock-in mechanism as presented in Sec. 4.3.2 perfectly enables fast restart of the clock generator when re-activating the core with minimized timing penalty.

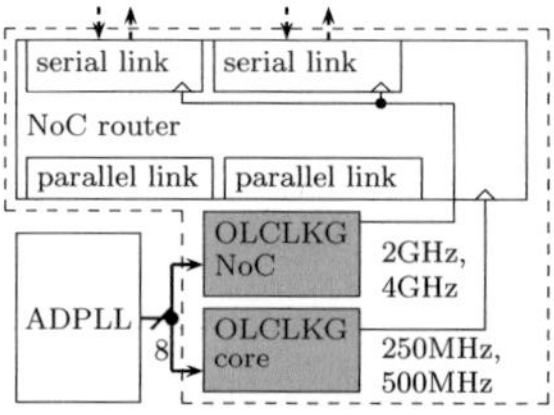

(a) clocking of a NoC router with multiple serial and parallel point-to-point links, adaptivity by frequency scaling based on the router bandwidth demands.

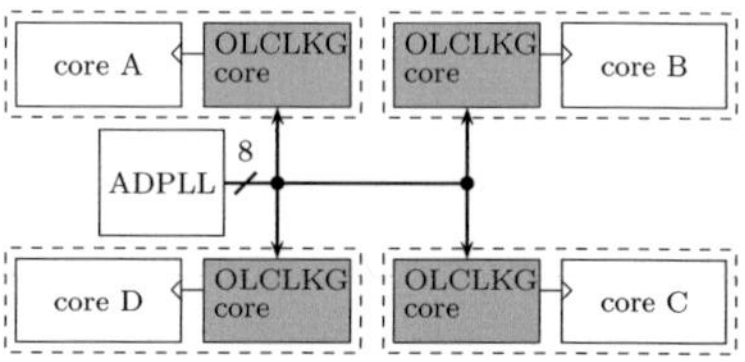

(b) clocking of four closely placed processor cores, with individual core frequency scaling capability

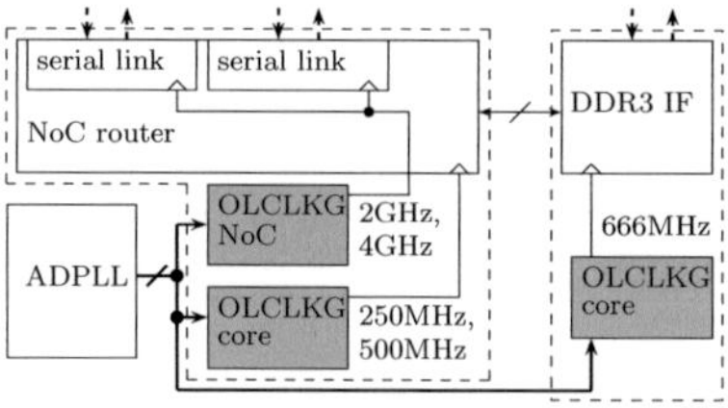

(c) clocking of a DDR3 RAM interface with closely placed NoC router with serial links to distribute the RAM data over longer distances on chip

Figure 5.28: Example scenarios for the application of the ADPLL with open-loop clock generators within GALS MPSoCs, core wrappers visualized as dashed lines

5.3 Summary

Open-loop methods for frequency multiplication and division have been presented. They allow to generate a wide range of output frequencies from a fixed frequency input signal. This enables to create high frequency clock signals for high-speed NoC link clocking and lower frequency core clock signals. The output frequency of the open-loop clock generator can be changed instantaneously, because no time consuming re-lock in of a closed loop ADPLL is required. This makes the proposed open-loop methods ideally suited for realization of ultra-fast DVFS schemes within GALS MPSoCs. Therefore an integration concept for the MPSoC core wrapper has been presented.

The circuit has been implemented in 65nm and 28nm CMOS technology. Its functionality has been successfully verified by measurements. The sensitivity of the open-loop clock generator with respect to phase mismatch in the input signals has been analyzed theoretically and verified by statistical measurement results.

6 Summary and Outlook

6.1 Summary

In this work novel circuit solutions for clock generation in heterogeneous GALS MPSoCs have been researched. Based on the clocking requirements with special emphasis on advanced power management techniques and network-on-chip fabrics with high speed serial links, the specifications of versatile clock generators which can be instantiated per-core have been defined. This basically includes a wide frequency range, ultra-fast frequency switching times, low jitter, small area and low power consumption for minimized integration overhead.
ADPLLs have been chosen as suitable circuit architecture for local clock generation based on a global reference clock. DCOs with multi-phase clock outputs are their key component. The main design challenge of the digital tuning scheme is to achieve a wide tuning range for robustness with respect to PVT variations and a small tuning step size for low jitter. To resolve this trade-off a new active bias circuit which is able to compensate the temperature and supply voltage dependency of the oscillator core has been developed and implemented into a DCO in 65nm CMOS technology. Thereby a small tuning step size for low jitter during operation can be achieved with a small digital tuning word. This helps to reduce the circuit effort within the ADPLL controller. As alternative approach for nanometer CMOS technologies, a 28nm DCO has been designed which achieves wide tuning range at small step size by a highly digital tuning approach with switchable resistors in the supply path. This circuit perfectly scales with technology and thereby leads to a compact solution in this advanced CMOS node.
A theory on the phase error reduction behavior of differential clock buffers with cross-coupled inverters has been presented. These circuits are essential for distribution of multi-phase clock signals at high frequencies as required in this work and can help to reduce signal imperfectness due to device mismatch.
A simple BBADPLL architecture is chosen for closed loop control of the DCOs because this minimalistic control scheme based on a binary PFD does not require high circuit effort. A numerical model of the control loop has been developed which

is used for system simulations and prediction of the circuit performance with special focus on lock-in time and jitter. An ultra-compact 65nm CMOS ADPLL has been implemented and evaluated successfully by testchip measurements. The circuit is capable to meet the jitter requirements of DDR2 and DDR3 memory interfaces. For further improvement of the BBADPLL jitter performance, a new controller clocking scheme using the divided DCO clock is proposed which can reduce the control loop delay and thereby reduce jitter accumulation. This scheme is used in a 28nm CMOS version of the ADPLL. The initial ADPLL lock-in time can be reduced significantly by a novel single-shot phase synchronization scheme. This enables fast frequency detection with high accuracy for frequency lock and allows to start the ADPLL near to its phase lock condition, thereby achieving instantaneous lock. The circuit has been analyzed theoretically and its fast lock-in capabilities have been measured successfully within a 28nm CMOS testchip.

A wide range of core clock frequencies can be generated from multi-phase DCO clock signals at a fixed frequency by open-loop methods. An open-loop clock generator which generates 50% duty cycle output clocks has been proposed and implemented in both 65nm and 28nm CMOS technology. This circuit allows instantaneous frequency changes which allow application of these clock generators for ultra fast DVFS. Also high-speed clock signals for advanced on-chip comminication circuits can be realized by the proposed clock generators. A theoretical analysis on the open-loop clock generator sensitivity to phase mismatch has been presented.

As result, solutions for ultra-compact ADPLL based clock generators for application in heterogeneous GALS MPSoCs have been developed. They employ various novel circuit techniques to improve the key performances for MPSoC target application. This especially includes lock-in time reduction, low jitter operation, robustness with respect to PVT variations and instantaneous frequency changes. Thereby the circuits are compact in terms of chip area and energy efficient. Thus they are suitable for a wide range of clocking applications for MPSoC cores, NoC links and I/O, helping to reduce the design implementation overhead by component re-use for future MPSoCs. The circuits have been verified successfully by measurements of three testchips in 65nm and 28nm CMOS technology, where they are in operational use to clock other system components.

6.2 Clock Generator Application

The circuits that have been developed in this work are applied in different research testchips at Technische Universität Dresden. Herein they serve for different clocking purposes which highlights both their flexibility and the capability of application as IP core. Two examples are briefly shown in the following.

"Tomahawk2", TSMC 65nm LP CMOS The "Tomahawk 2" testchip is a complex MPSoC demonstrator developed within the CoolBaseStations project. It logic architecture is developed at the VODAFONE Chair of Mobile Communication Systems[1] and its infrastructure circuits for clocking power management and on-chip communication are developed by the Chair of Highly-Parallel VLSI-Systems and Neuromorphic Circuits[2] which also is responsible for backend design implementation. The top level floor plan diagram is shown in Fig. 6.1. It includes 12 processing cores, a DDR2 interface and an LVDS interface for communication with an FPGA. The cores are embedded into the wrapper topology including components for clocking, power management and NoC communication as shown in Sec. 2.5. The cores are partly enabled for ultra-fast DVFS with combined AVFS functionality. The PL changes are controlled by a hardware assisted task scheduling system. A packed based NoC connects the cores and employs high-speed serial on-chip links (see Sec. 2.4) for the long distance point to point connections. This allows to realize a compact floorplan as shown in Fig. 6.1. The GALS clocking architecture is driven by 18 ADPLLs as described in Sec. 4.2. With "Tomahawk 2" the scalability of the heterogeneous MPSoC infrastructure circuits developed at the Chair of Highly-Parallel VLSI-Systems and Neuromorphic Circuits, including the clock generators from this work, to more complex chips shall be proven[3].

"Titan", GLOBALFOUNDRIES 28nm SLP CMOS The "Titan" testchip as shown in Fig. 6.2 is a heterogeneous system which has been developed and implemented by the Chair of Highly-Parallel VLSI-Systems and Neuromorphic Circuits in an advanced 28nm CMOS technology. It includes cores with various functionality, like for example neuromorphic mixed-signal circuits, analog-to-digital converters (ADCs), test circuits for IR-drop analysis within SoCs, high-speed serial I/O circuits, a transceiver for 3D chip stack communication over through-silicon vias (TSVs) and test structures for high-speed serial on-chip links as shown in Sec. 2.4. All functional cores within its core wrappers are clocked by the 28nm ADPLL circuit from this work as presented in Sec. 4.3. Herein flexibility is achieved by using the multi-phase clock outputs of the DCO as shown in Sec. 3.3 for various applications. The clock data recovery circuit of the high-speed serial I/O employs

[1]https://mns.ifn.et.tu-dresden.de
[2]http://hpsn.et.tu-dresden.de
[3]Tape-out planned for 03/2013

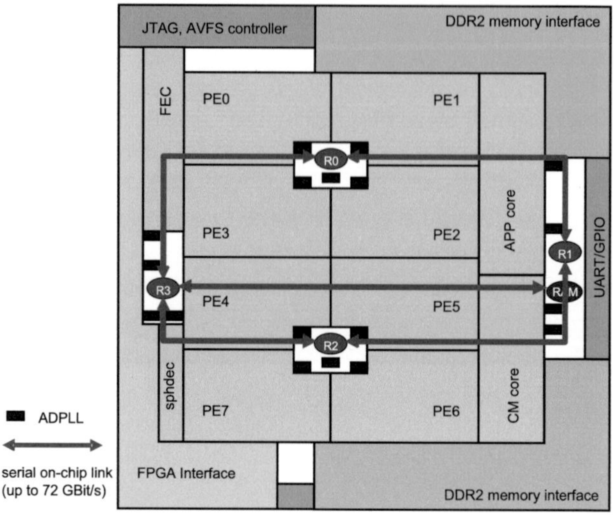

Figure 6.1: "Tomahawk2" block level floorplan, 6mm×6mm, 65nm CMOS, modified from [Eis12]

the multi-phase DCO clock signals for oversampling operation. Within the IR-drop analysis and measurement circuits eight clock phases at 500ps nominal period are employ to realize programmable skews of measurement clocks with 62.5ps timing resolution. Moreover the high-speed clocks which are generated by a multi-phase clock multiplier as shown in Sec. 5.1.1 of this work are used to drive high-speed serial links for both on-chip and 3D chip stack communication.

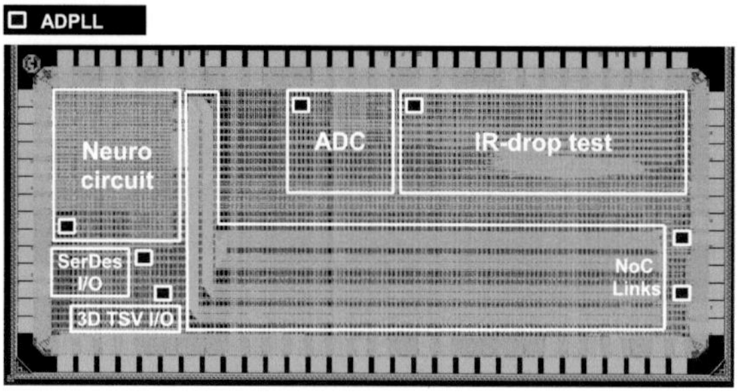

Figure 6.2: Layout of the "Titan" testchip, 3.0mm × 1.5mm, 28nm CMOS

6.3 Further Work

The clock generation circuits that have been developed in this work can be improved and extended in further work. Some aspects are summarized in the following.

The noise which is generated by the active current bias generator for supply voltage and temperature compensation of the DCO in 65nm CMOS technology (see Sec. 3.2) can be reduced in a re-design to achieve the same low jitter performance as the uncompensated version. This can be done using additional filter capacitors within the DC bias circuits at cost of chip area or higher power consumption in the bias circuits itself. This trade-off with respect to the overall ADPLL performance in terms of power consumption and chip area must be evaluated carefully.

The circuit architectures of the DCOs and ADPLLs can be reviewed to identify suitable design parameters as "tweaking knobs" that allow do adapt these circuits at design time to a more specific clocking application. Thereby individual solutions for specific design targets (e.g. ultra low-power with relaxed jitter constraints, ultra-low jitter) can be realized to improve the overhead of the clocking circuits within the MPSoC in terms of chip area and power consumption. However, in order to allow flexible re-use and efficient silicon implementation, the tweaking of these parameters should be highly automated. As example, in case of the 28nm DCO realization as presented in Sec. 3.3 this can be achieved by replacing the oscillator core within the supply voltage tuning scheme. High drive strength cells can be used for low jitter design targets and low drive strength cells can be used for ultra-low power consumption. Also different architectures (e.g. single-ended) are possible if no multi-phase clocks are required. Also the digital part of the ADPLL can be configurable to adjust the loop filter accuracy (in terms of bits per word) for different target applications.

For application of multiple ADPLLs within a single MPSoC strategies for built-in self-test are required to reduce production test time. This includes autonomous, automated measurement of the key performances of the ADPLLs as for example lock-in time and output clock frequency and can also consider advanced clock quality measures such as jitter. In a further step these on-chip clock measurement circuits can be used to calibrate the ADPLLs to an optimal design point in a built in calibration scheme. As example the loop filter coefficients α and β could be adjusted during system runtime to ensure operation at the point of minimum accumulated jitter as indicated by the analysis results in Sec. 4.1.4 of this work. Thereby runtime variations of circuit parameters (e.g. DCO gain K_t) due to temperature drifts could be tracked. These on-chip measurement and calibration circuits should be ultra-

compact for low area overhead. Ideally they are located centrally and are re-used for multiple ADPLLs on the same chip. This also includes the demand to centralize configuration effort of multiple ADPLLs, for example by using low speed serial on-chip interfaces for configuration. This helps to reduce the complexity of the JTAG custom test logic which in this work is individual for each ADPLL instance. Thereby the gate count of the MPSoC core wrapper logic can be further reduced.
Due to the mainly digital nature of the clock generation circuits of this work, they are well suited for easy design migration to the next CMOS technology nodes, as for example 22nm [AAB+12] or 14nm [War11]. Here a general reduction of the custom design content is desired to allow implementation using highly automated digital implementation flows, that are able to handle complex physical design rules including design for manufacturability (DFM), design for yield (DFY) and reliability aspects [War11].
As result, the approaches of this work have the potential to be extended towards flexible clocking IP generators for advanced CMOS technology nodes.

A Appendix

A.1 Jitter Definitions

Jitter describes the uncertainty and fluctuation of timing events in electronic signals [HWB04] and is a commonly used measure for the purity of clock signals. Based on this general definition, a wide range of concrete jitter measures exist, which allow signal quality description based on different target applications (e.g. digital system clocking, wire line communication, wireless signal transmission). In the following some relevant jitter definitions for MPSoC clocking applications as used in this work are summarized. They are derived from [Kun05], [Lee02] and [MR09].
It is $\{t_k\}$ a sequence of times at which positive clock edges occur, as illustrated in Fig. A.1. T_k is the period of clock cycle k

$$T_k = t_{k+1} - t_k \tag{A.1}$$

and the average period of the clock signal is

$$T = \overline{T_k}. \tag{A.2}$$

The *absolute jitter* is defined as the sequence

$$\{j_{t,\text{abs}}(k)\} = \{t_k - k \cdot T\} = \{\sum_{i=1}^{k}(T_i - T)\}. \tag{A.3}$$

Its standard deviation is

$$\sigma_{t,\text{abs}}^2 = \text{Var}(j_{t,\text{abs}}) = \overline{j_{t,\text{abs}}^2} - \overline{j_{t,\text{abs}}}^2. \tag{A.4}$$

The absolute jitter is a critical measure for systems where different clock sources have to be synchronized [vdTKvR03]. It describes the long term accuracy of clock signals. In the BBADPLL model as presented in Sec. 4.1.2 of this work the absolute jitter can be used to describe the standard deviation of the PFD input timing difference $(t_{\text{ref}} - t_{\text{div}})$, when operating in a DCO jitter dominated regime where

$\sigma^2_{\mathrm{T,DCO}} \gg \sigma^2_{T,ref}$.

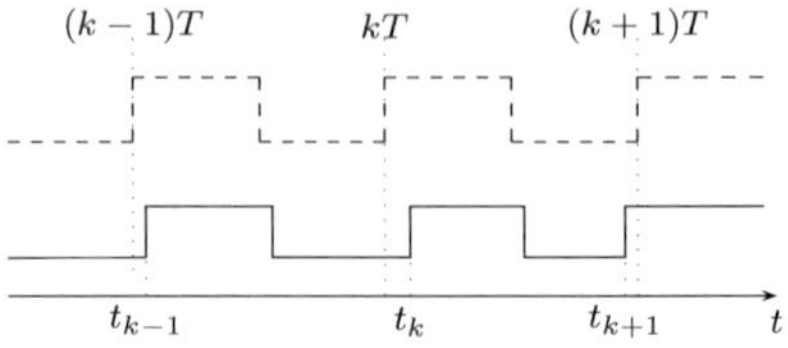

Figure A.1: Jitter definition

The *accumulated jitter* over n clock cycles (also called n *cycle jitter*) is defined as the time difference between two clock edges with distance n, which is the sum of n adjacent clock periods T_k with respect to its average value $n \cdot T$

$$\{j_{T,\mathrm{acc},n}(k)\} = \{t_{k+n} - t_k - n \cdot T\}. \tag{A.5}$$

This jitter metric is self referenced, since it does not depend on an absolute time. Its variance is

$$\sigma^2_{T,\mathrm{acc},n} = \mathrm{Var}(j_{T,\mathrm{acc},n}) = \overline{j^2_{T,\mathrm{acc},n}} - \overline{j_{T,\mathrm{acc},n}}^2. \tag{A.6}$$

The accumulated jitter describes the timing relation between n adjacent clock edges, which is an important metric for clock data recovery applications, where clock edges are locked to a data stream and must keep their ideal sampling position over n cycles until the clock phase is updated again. Also the clock signal specification for DDR2 and DDR3 memory interfaces defines constraints for the accumulated jitter. Fig. A.2 illustrates jitter accumulation which leads to increased variation of the clock edge timing after n cycles.

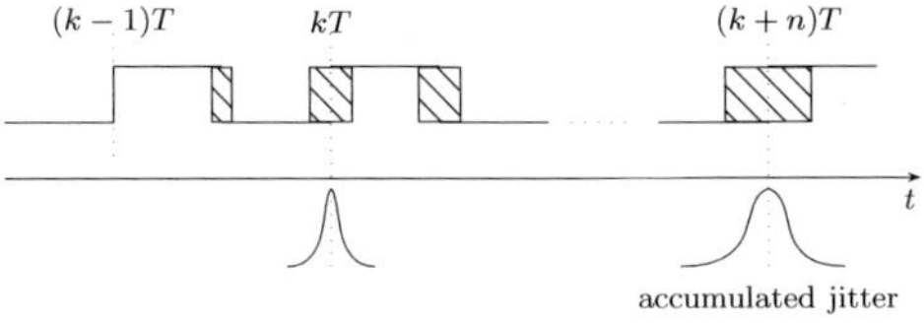

Figure A.2: Accumulated jitter

Period jitter is a special case of accumulated jitter for $n = 1$. It is defined as

$$\{j_T(k)\} = \{t_{k+1} - t_k - T\} = \{T_k - T\}. \tag{A.7}$$

Therefore the period jitter is also called *cycle jitter* [HWB04]. Its variance reads

$$\sigma_T^2 = \mathrm{Var}(j_T) = \overline{j_T^2} - \overline{j_T}^2. \tag{A.8}$$

In digital systems the period jitter is the main jitter measure describing the clock uncertainty with respect logic speed. In clocked digital sequential circuits data is launched at a first sequential element (e.g. flip flop) at t_k and is received at t_{k+1}. The period jitter thus directly reduces the setup time margin and thereby limits the maximum operating frequency [Lee02].

The *cycle-to-cycle jitter* [HWB04] is defined as the difference of two adjacent clock periods

$$\{j_{cc}\} = \{T_{k+1} - T_k\}. \tag{A.9}$$

The variance of the cycle-to-cycle jitter is

$$\sigma_{cc}^2 = \mathrm{Var}(j_{cc}) = \overline{j_{cc}^2} - \overline{j_{cc}}^2. \tag{A.10}$$

Cycle-to-cycle jitter is part of the DDR2 and DDR3 memory interface clock specification.

For the ADPLL analysis in this work the jitter accumulation is expressed by both *absolute jitter*, describing the closed loop noise behavior of the ADPLL and accumulated jitter, which can be measured in the manufactured chips using a sampling oscilloscope [MR09].

When operating in a DCO noise dominated regime where $\sigma_{T,DCO}^2 \gg \sigma_{T,ref}^2$ the reference clock jitter can be neglected. In this case the accumulated jitter variance $\sigma_{T,\mathrm{acc},n}$ for $n \to \infty$ is finite [MR09] since the closed loop ADPLL adjusts the timing of the DCO such that its phase follows the reference clock signal phase for fluctuations well below the loop filter bandwidth (e.g. at DC frequency).

Due to the fact that the accumulated jitter of the (divided) DCO signal is self referenced, whereas the absolute jitter is defined as timing difference between the (divided) DCO signal edges and the ideal reference clock. As illustrated in Fig. A.3 it is with Eq. A.3 and Eq. A.5

$$\{j_{T,\mathrm{acc},n}(k)\} = \{j_{t,\mathrm{abs}}(k+n)\} - \{j_{t,\mathrm{abs}}(k)\}. \tag{A.11}$$

For large values of n the correlation between the absolute jitter values at $j_{t,\mathrm{abs}}(k)$ and $j_{t,\mathrm{abs}}(k+n)$ is negligible. The noise transfer function between of the DCO noise to the output node has high-pass characteristics with a gain of $|H_{\text{noise}}|(0) \to -\infty$ at DC frequencies [MR09] as illustrated in the description of the ADPLL noise shaping behavior in Sec. 4.1.2. Therefore its impulse response converges to $\text{IFFT}(H_{\text{noise}})(\infty) \to 0$ for large times, i.e. values of n. The ADPLL is capable to suppress the absolute jitter at the normalized time k until $k+n$ is reached for large n. Therefore the variance of the accumulated jitter, as difference of two uncorrelated absolute jitter values, can be approximated by

$$\sigma^2_{T,\text{acc},n\to\infty} = 2 \cdot \sigma^2_{t,\text{abs}} = 2 \cdot \sigma^2_{\Delta t,\text{PFD}}. \tag{A.12}$$

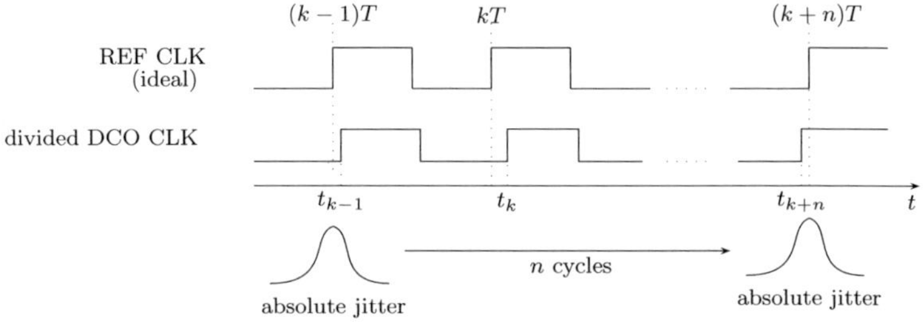

Figure A.3: Relation between absolute and accumulated jitter for ADPLLs with ideal reference clock

Besides these time domain specifications, the jitter phenomenon can be analyzed in the frequency domain [HHS08]. This relates to the spectral purity of the clock signals and are important in applications for wireless data transmission [Höp08]. These analyses can also be useful when performing system analyses of closed regulation loops in frequency domain as for example presented in [DD06, DD08, ZTL+09] for the analysis of BBADPLLs.

A.2 EDA Tools Used in this Work

Table A.1: Design tools overview

purpose	**tool**
Schematic entry	Cadence® Virtuoso™
Layout entry	Cadence® Virtuoso™
Layout verification and parasitic extraction	Mentor® Calibre™
Analog circuit simulation	Cadence® Spectre™
Analog circuit verification and optimization	MunEDA® WiCkeD™
Digital circuit simulation	Cadence® NCSim™
Mixed-signal circuit simulation	Cadence® AMS designer™
Logic synthesis	Synopsys® DesignCompiler™
Place&Route	Synopsys® ICCompiler™
STA	Synopsys® PrimeTime™
System modeling	MATLAB™
System modeling	GNU Octave

A.3 Measurement Setups

The implemented ADPLL circuits as presented in Sec. 4.2 and Sec. 4.3 feature various debug and measurement functionality which can be controlled via a JTAG interface. This includes:

- Readout of ADPLL controller status information (e.g. current tuning word, lock status, tuning overflow flags).
- Configuration of the loop filter parameters α and β.
- Configuration of the lock-in detection timing window.
- Configuration (including disable) of the fast-lock in functionality of the 28nm ADPLL as shown in Sec. 4.3.2.
- Configuration of DCO bias settings for active supply voltage and temperature compensation and fine tune gain, in case of the 65nm DCO and the number of always-on tuning switches in case of the 28nm DCO.
- Operation of the DCO in open-loop mode with direct definition of the digital tuning word. In this mode the lock detection is disabled and the lock bit is enforced to 1 to open all output clock gates of the ADPLL clock generators.

The output signals of all ADPLL clock generators on the testchips as presented in Sec. 2.8 can be fed to standard digital low-speed I/O pads which are capable to transmit signals up to $\approx$ 200MHz. For high speed signal measurements at least one ADPLL per testchip is connected to a high-speed LVDS pad which is capable to transmit clock signals with frequencies higher than 2GHz. Both the NoC clock output and the core clock output can be multiplexed to the LVDS pad, where the first option is used for DCO measurements and the second one is employed for evaluation of the open-loop clock generator. Each ADPLL has a reference bypass mode where the reference clock can be directly routed to the output.
Fig. A.4 shows the principle measurement setup that has been used for characterization of the testchips "Tommy", "Atlas" and "Cool28SoC" (with some slight modifications).
The testchips are mounted on a power supply printed circuit board (PCB) which have been designed by the Chair of Highly-Parallel VLSI-Systems and Neuromorphic Circuits. Fig. A.5 shows the PCB setups for the three testchips measured in this work. The main supply voltages are generated by PMICs which are controllable

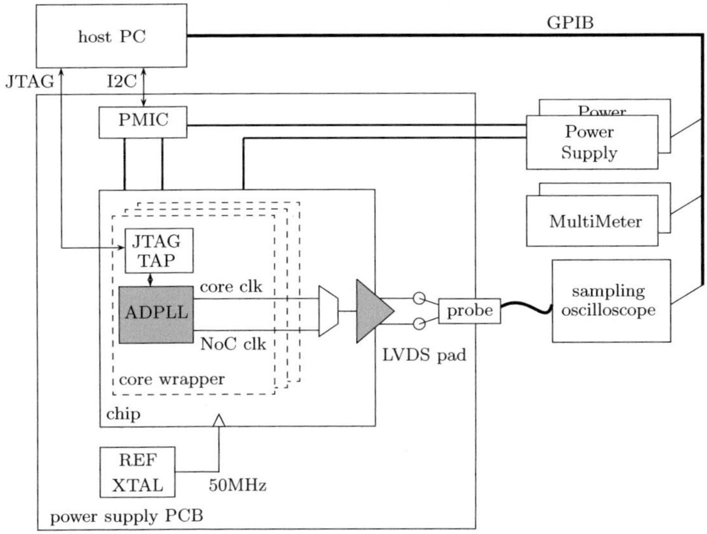

Figure A.4: Measurement setup

from the host PC. Selected supply voltages for the testchips can be provided directly from measurement power supplies (e.g. DCO supply voltages). On the PCB the reference clock is generated. Connectivity to the host PC (via JTAG) and to additional FPGA boards is provided. The clock output LVDS pads are probed using a high-speed differential probe device of a digital sampling oscilloscope LeCroy WavePro 7300a, which allows clock timing measurements with a jitter noise floor of 3ps rms. Various multimeters allow current and voltage measurements of the system, e.g. to determine the power consumption of the chip components. The thermal measurements described in Sec. 3.2 have been executed on "Atlas" by using a Peltier Device for active heating and cooling of the testchip.
The measurement equipment is controlled from the host PC over GPIB. This allows to implement complex measurement tasks as MATLAB scripts which run automatically and capture results.
To estimate the accuracy of the jitter measurements in this work, the measurement environment must be characterized. The reference clock signal is the time base for the ADPLL clock generators, as analyzed in Sec. 4.1.4 its jitter influences the output jitter of the clock generators. Within the test setups the reference clock jitter is measured by feeding it through the ADPLL clock generators to the LVDS measurement pad using the bypass mode. Tab. A.2 summarizes the measured

(a) "Atlas"

(b) "Tommy"

(c) "Cool28SoC"

Figure A.5: Testchip PCB photos

reference clock jitter properties for the three testchip boards.

Table A.2: Reference clock jitter measured through on-chip bypass

testchip	T_{ref} [ns]	$\sigma_{T,\text{ref}}$ [ps]	$\sigma_{acc,\infty,\text{ref}}$ [ps]
"Tommy"	20	7.9	15.7
"Atlas"	20	8.2	16.5
"Cool28SoC"	20	7.5	11.8

In case of the 28nm ADPLL realization on "Cool28SoC" the internal jitter values are small compared to the accuracy and noise floor of the on-chip frequency dividers, LVDS drivers and the measurement setup components. This is illustrated in Fig. A.6, where the accumulated jitter $\sigma_{T,\text{acc},n}$ of the 28nm ADPLL is plotted. Here accumulation is performed using the internal open-loop clock generator as presented in Sec. 5.1 as frequency divider. The accumulation values for $n = 1$ and $n = 2$ are realized by direct output of the DCO to the LVDS pad and a single by 2 frequency divider. All these components add jitter, which lead to the fact that the accumulated jitter does not increase with $\propto \sqrt{n}$ as it would be expected for addition of noisy DCO periods with domiating white noise jitter [MR09]. Note that here the number of accumulated DCO cycles is small compared to the closed loop regulation behavior of the ADPLL because the loop divider is $N = 40$. The expected jitter accumulation characteristics is observed for $n > 7$. In this region a curve is fitted to the measurement data and extrapolated to the root DCO period jitter of $\sigma_{T,\text{DCO}} \approx 3$ps. The noise floor of the on-chip and off-chip frequency divider and buffer components is $\approx$ 8ps rms.

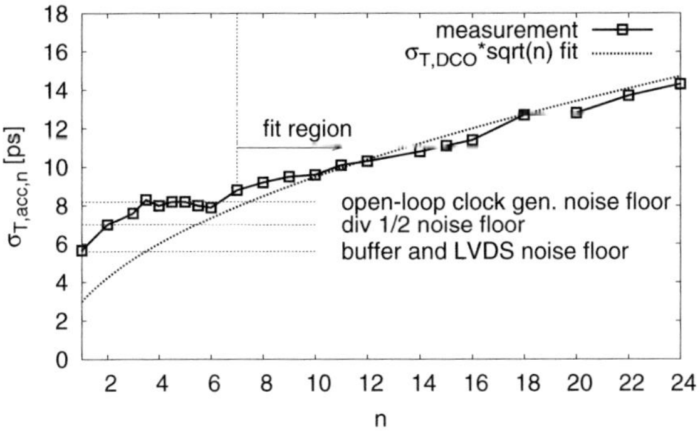

Figure A.6: 28nm ADPLL output jitter accumulated over few clock cycles

Publications

[BNS+11] Volker Boos, Jacek Nowak, Matthias Sylvester, Stephan Henker, Sebastian Höppner, Heiko Grimm, Dominik Krausse, and Ralf Sommer. Strategies for initial sizing and operating point analysis of analog circuits. In *Design, Automation Test in Europe Conference Exhibition (DATE), 2011*, pages 1 –3, Mar. 2011.

[EMF+12] F. Ellinger, T. Mikolajik, G. Fettweis, D. Hentschel, S. Kolodinski, H. Warnecke, T. Reppe, C. Tzschoppe, J. Dohl, C. Carta, D. Fritsche, M. Wiatr, S.D. Kronholz, R.P. Mikalo, H. Heinrich, R. Paulo, R. Wolf, J. Hubner, J. Waltsgott, K. Meissner, R. Richter, M. Bausinger, H. Mehlich, M. Hahmann, H. Moller, M. Wiemer, H.-J. Holland, R. Gartner, S. Schubert, A. Richter, A. Strobel, A. Fehske, S. Cech, U. Assmann, S. Höppner, D. Walter, H. Eisenreich, and R. Schüffny. Cool silicon ICT energy efficiency enhancements. In *Semiconductor Conference Dresden-Grenoble (ISCDG), 2012 International*, pages 1 –4, Sep. 2012.

[EMF+13] F. Ellinger, T. Mikolajick, G. Fettweis, D. Hentschel, S. Kolodinski, H. Warnecke, T. Reppe, C. Tzschoppe, J. Dohl, C. Carta, D. Fritsche, G. Tretter, M. Wiatr, S.D. Kronholz, R.P. Mikalo, H. Heinrich, R. Paulo, R. Wolf, J. Hübner, J. Waltsgott, K. Meißner, R. Richter, O. Michler, M. Bausinger, H. Mehlich, M. Hahmann, H. Möller, M. Wiemer, H.-J. Holland, R. Gärtner, S. Schubert, A. Richter, A. Strobel, A. Fehske, S. Cech, U. Aßmann, A. Pawlak, M. Schröter, W. Finger, S. Schumann, S. Höppner, D. Walter, H. Eisenreich, and R. Schüffny. Energy efficiency enhancements for semiconductors, communications, sensors and software achieved in Cool Silicon cluster project. *European Journal on Applied Physics (EPJAP)*, 63, 2013.

[GHH+11] Johannes Görner, Sebastian Höppner, Stephan Henker, Rene Schüffny, and Achim Graupner. A matrix-based voltage range estimation method using linearized operating points. In *Mixed Design of Integrated Circuits and Systems (MIXDES), 2011 Proceedings of the 18th International Conference*, pages 422 –427, Jun. 2011.

[HEH+13] S. Höppner, H. Eisenreich, S. Henker, D. Walter, G. Ellguth, and R. Schüffny. A compact clock generator for heterogeneous GALS MPSoCs in 65-nm CMOS technology. *Very Large Scale Integration (VLSI) Systems, IEEE Transactions on*, 21(3):566 –570, 2013.

[HGH+10] S. Höppner, J. Görner, S. Henker, R. Schüffny, and A. Graupner. A lookup table flow for analog design automation. In *Proceedings of GMM/ITG-Fachtagung Analog 2010*, pages 179–184. VDE Verlag, Mar. 2010.

[HHES11] Sebastian Höppner, Stephan Henker, Holger Eisenreich, and Rene Schüffny. An open-loop clock generator for fast frequency scaling in 65nm CMOS technology. In *Mixed Design of Integrated Circuits and Systems (MIXDES), 2011 Proceedings of the 18th International Conference*, pages 264 –269, Jun. 2011.

[HHH+12] Sebastian Höppner, Stefan Haenzsche, Stephan Hartmann, Stefan Schiefer, and Rene Schüffny. Temperature and supply voltage compensated biasing for digitally controlled oscillators. In *Mixed Design of Integrated Circuits and Systems (MIXDES), 2012 Proceedings of the 19th International Conference*, May 2012.

[HHS08] S. Höppner, S Henker, and R. Schüffny. A behavioral PLL model with timing jitter due to white and flicker noise sources. In *Proceedings of GMM/ITG-Fachtagung Analog 2008*, pages 125–130, Apr. 2008.

[HHSG10] Sebastian Höppner, Stephan Henker, Rene Schüffny, and Achim Graupner. A fast method for transistor circuit voltage range analysis using linear programming. In *Mixed Design of Integrated Circuits and Systems (MIXDES), 2010 Proceedings of the 17th International Conference*, pages 385 –390, Jun. 2010.

[HSE+12] S. Höppner, Chenming Shao, H. Eisenreich, G. Ellguth, M. Ander, and R. Schüffny. A power management architecture for fast per-core DVFS in heterogeneous MPSoCs. In *Circuits and Systems (ISCAS), 2012 IEEE International Symposium on*, pages 261 –264, May 2012.

[HSN09] S. Höppner, R. Schüffny, and M. Nemes. A low-power, robust multi-modulus frequency divider for automotive radio applications. In *Mixed Design of Integrated Circuits Systems, 2009. MIXDES '09. MIXDES-16th International Conference*, pages 205 –209, Jun. 2009.

[HSTW10] S. Höppner, R. Schüffny, Zuo-Min Tsai, and Huei Wang. Wide swing signal amplification by SC voltage doubling. In *Circuits and Systems (ISCAS), Proceedings of 2010 IEEE International Symposium on*, pages 761 –764, May 2010.

[HWES10] S. Höppner, D. Walter, H. Eisenreich, and R. Schüffny. Efficient compensation of delay variations in high-speed network-on-chip data links. In *System on Chip (SoC), 2010 International Symposium on*, pages 55 –58, Sep. 2010.

[HWES11] S. Höppner, D. Walter, G. Ellguth, and R. Schüffny. Mismatch characterization of high-speed NoC links using asynchronous sub-sampling. In *System on Chip (SoC), 2011 International Symposium on*, Sep. 2011.

[HWES12] Sebastian Höppner, Dennis Walter, Georg Ellguth, and René Schüffny. On-chip measurement and compensation of timing imbalances in high-speed serial NoC links. *International Journal of Embedded and Real-Time Communication Systems (IJERTCS)*, 3(4):42–56, 2012.

[SEH+12] Stefan Scholze, Holger Eisenreich, Sebastian Höppner, Georg Ellguth, Stephan Henker, Mario Ander, Stefan Hänzsche, Johannes Partzsch, Christian Mayr, and Rene Schüffny. A 32 GBit/s communication SoC for a wafer-scale neuromorphic system. *Integration, the VLSI Journal*, 45(1):61 – 75, 2012.

[SSP+11] Stefan Scholze, Stefan Schiefer, Johannes Partzsch, Stephan Hartmann, Christian Georg Mayr, Sebastian Höppner, Holger Eisenreich, Stephan Henker, Bernhard Vogginger, and Rene Schüffny. VLSI implementation of a 2.8 gevent/s packet based AER interface with routing and event sorting functionality. *Frontiers in Neuroscience*, 5(00117), 2011.

[UHES10] J. Uhlig, S. Höppner, G. Ellguth, and R. Schüffny. A low-power cell-based-design multi-port register file in 65nm CMOS technology. In *Circuits and Systems (ISCAS), Proceedings of 2010 IEEE International Symposium on*, pages 313 –316, May 2010.

[WHE+12] D. Walter, S. Höppner, H. Eisenreich, G. Ellguth, S. Henker, S. Haenzsche, R. Schüffny, M. Winter, and G. Fettweis. A source-synchronous 90Gbit/s capacitively driven serial on-chip link over 6mm in 65nm CMOS. In *Solid-State Circuits Conference, 2012. ISSCC 2012. Digest of Technical Papers. IEEE International*, Feb. 2012.

[WKA+12] M. Winter, S. Kunze, E.P. Adeva, B. Mennenga, E. Matus, G. Fettweis, H. Eisenreich, G. Ellguth, S. Höppner, S. Scholze, R. Schüffny, and T. Kobori. A 335Mb/s 3.9mm2 65nm CMOS flexible MIMO detection-decoding engine achieving 4G wireless data rates. In *Solid-State Circuits Conference Digest of Technical Papers (ISSCC), 2012 IEEE International*, pages 216 –218, Feb. 2012.

References

[AAB+12] C. Auth, C. Allen, A. Blattner, D. Bergstrom, M. Brazier, M. Bost, M. Buehler, V. Chikarmane, T. Ghani, T. Glassman, R. Grover, W. Han, D. Hanken, M. Hattendorf, P. Hentges, R. Heussner, J. Hicks, D. Ingerly, P. Jain, S. Jaloviar, R. James, D. Jones, J. Jopling, S. Joshi, C. Kenyon, H. Liu, R. McFadden, B. McIntyre, J. Neirynck, C. Parker, L. Pipes, I. Post, S. Pradhan, M. Prince, S. Ramey, T. Reynolds, J. Roesler, J. Sandford, J. Seiple, P. Smith, C. Thomas, D. Towner, T. Troeger, C. Weber, P. Yashar, K. Zawadzki, and K. Mistry. A 22nm high performance and low-power CMOS technology featuring fully-depleted tri-gate transistors, self-aligned contacts and high density MIM capacitors. In *VLSI Technology (VLSIT), 2012 Symposium on*, pages 131 –132, Jun. 2012.

[ABR+99] P. Andreani, F. Bigongiari, R. Roncella, R. Saletti, and P. Terreni. A digitally controlled shunt capacitor CMOS delay line. *Analog Integrated Circuits and Signal Processing*, 18(1):89 –96, Jan. 1999.

[AF10] O. Arnold and G. Fettweis. Power aware heterogeneous MPSoC with dynamic task scheduling and increased data locality for multiple applications. In *Embedded Computer Systems (SAMOS), 2010 International Conference on*, pages 110 –117, Jul. 2010.

[AF11] O. Arnold and G. Fettweis. Self-aware heterogeneous MPSoC with dynamic task scheduling for battery lifetime extension. In *Computing in Heterogeneous, Autonomous 'N' Goal-Oriented Environments (CHANGE), 2011 1st International Workshop on*, pages 1 –7, Mar. 2011.

[AJ04] Hannu Tenhunen Axel Jantsch. *Networks on Chip*. Kluwer Academic Publishers, 2004.

[ARK07] M.A. Abas, G. Russell, and D.J. Kinniment. Built-in time measurement circuits – a comparative design study. *Computers Digital Techniques, IET*, 1(2):87 –97, Mar. 2007.

[ATE+09] F. Arnaud, A. Thean, M. Eller, M. Lipinski, Y.W. Teh, M. Ostermayr, K. Kang, N.S. Kim, K. Ohuchi, J.-P. Han, D.R. Nair, J. Lian, S. Uchimura,

S. Kohler, S. Miyaki, P. Ferreira, J.-H. Park, M. Hamaguchi, K. Miyashita, R. Augur, Q. Zhang, K. Strahrenberg, S. ElGhouli, J. Bonnouvrier, F. Matsuoka, R. Lindsay, J. Sudijono, F.S. Johnson, J.H. Ku, M. Sekine, A. Steegen, and R. Sampson. Competitive and cost effective high-k based 28nm cmos technology for low power applications. In *Electron Devices Meeting (IEDM), 2009 IEEE International*, pages 1 –4, Dec. 2009.

[AWZ08] M. Ali, M. Welzl, and M. Zwicknagl. Networks on chips: Scalable interconnects for future systems on chips. In *Circuits and Systems for Communications, 2008. ECCSC 2008. 4th European Conference on*, pages 240–245, Jul. 2008.

[Bak05] R. Jacob Baker. *CMOS Circuit Design, Layout and Simulation, Second Edition.* IEEE, WILEY-INTERSCIENCE, 2005.

[Bha09] A. B. Bhattacharyya. *Compact MOSFET Models for VLSI Design.* WILEY, 2009.

[BMM07] A. Banerjee, R. Mullins, and S. Moore. A power and energy exploration of network-on-chip architectures. In *Networks-on-Chip, 2007. NOCS 2007. First International Symposium on*, pages 163 –172, May 2007.

[Bra99] N. Bray. Designing for the ip supermarket. In *Fall VIUF Workshop, 1999.*, pages 8 –13, Oct. 1999.

[BSS+08] F. Barale, P. Sen, S. Sarkar, S. Pinel, and J. Laskar. Programmable frequency-divider for millimeter-wave PLL frequency synthesizers. In *Microwave Conference, 2008. EuMC 2008. 38th European*, pages 460 –463, Oct. 2008.

[CC08] Y.A. Chau and C.-F. Chen. High-performance glitch-free digital frequency synthesiser. *Electronics Letters*, 44(18):1063 –1064, 2008.

[CCL05] Pao-Lung Chen, Ching-Che Chung, and Chen-Yi Lee. A portable digitally controlled oscillator using novel varactors. *Circuits and Systems II: Express Briefs, IEEE Transactions on*, 52(5):233 – 237, May 2005.

[CCYL06] Pao-Lung Chen, Ching-Che Chung, Jyh-Neng Yang, and Chen-Yi Lee. A clock generator with cascaded dynamic frequency counting loops for wide multiplication range applications. *Solid-State Circuits, IEEE Journal of*, 41(6):1275–1285, Jun. 2006.

[CL09] J.-Y. Chang and S.-I. Liu. A 1.5 GHz phase-locked loop with leakage current suppression in 65 nm CMOS. *Circuits, Devices Systems, IET*, 3(6):350 – 358, Dec. 2009.

[CL10] Jung-Yu Chang and Shen-Iuan Liu. A phase-locked loop with background leakage current compensation. *Circuits and Systems II: Express Briefs, IEEE Transactions on*, 57(9):666 –670, Sep. 2010.

[CS96] J. Craninckx and M. S. J. Steyaert. A 1.75-GHz/3-V dual-modulus divide-by-128/129 prescaler in 0.7-μm CMOS. *IEEE Journal of Solid-State Circuits*, 31(7):890–897, Jul. 1996.

[CSM10] M.S.-W. Chen, D. Su, and S. Mehta. A calibration-free 800 MHz fractional-N digital PLL with embedded TDC. *Solid-State Circuits, IEEE Journal of*, 45(12):2819 –2827, Dec. 2010.

[CYL09] Man-Chia Chen, Jui-Yuan Yu, and Chen-Yi Lee. A sub-100 uw area-efficient digitally-controlled oscillator based on hysteresis delay cell topologies. In *Solid-State Circuits Conference, 2009. A-SSCC 2009. IEEE Asian*, pages 89 –92, Nov. 2009.

[DC04] Kurt Keutzer David Chinnery. *Closing the Gap Between ASIC & Custom.* Kluwer Academic Publishers, 2004.

[DD05] N. Da Dalt. A design-oriented study of the nonlinear dynamics of digital bang-bang PLLs. *Circuits and Systems I: Regular Papers, IEEE Transactions on*, 52(1):21–31, Jan. 2005.

[DD06] N. Da Dalt. Markov chains-based derivation of the phase detector gain in bang-bang PLLs. *Circuits and Systems II: Express Briefs, IEEE Transactions on*, 53(11):1195 –1199, Nov. 2006.

[DD08] N. Da Dalt. Linearized analysis of a digital bang-bang PLL and its validity limits applied to jitter transfer and jitter generation. *Circuits and Systems I: Regular Papers, IEEE Transactions on*, 55(11):3663 –3675, Dec. 2008.

[DGJ+12] S. Damaraju, V. George, S. Jahagirdar, T. Khondker, R. Milstrey, S. Sarkar, S. Siers, I. Stolero, and A. Subbiah. A 22nm IA multi-CPU and GPU system-on-chip. In *Solid-State Circuits Conference Digest of Technical Papers (ISSCC), 2012 IEEE International*, pages 56 –57, Feb. 2012.

[DHCC03] Wei-Jin Dai, D. Huang, Chin-Chih Chang, and M. Courtoy. Silicon virtual prototyping: the new cockpit for nanometer chip design [SoC]. In *Design Automation Conference, 2003. Proceedings of the ASP-DAC 2003. Asia and South Pacific*, pages 635 – 639, Jan. 2003.

[DS05] Azadeh Davoodi and Ankur Srivastava. Wake-up protocols for controlling current surges in MTCMOS-based technology. In *Design Automation Conference, 2005. Proceedings of the ASP-DAC 2005. Asia and South Pacific*, volume 2, pages 868 – 871 Vol. 2, Jan. 2005.

[EFS96] G. Enrique Fernandez and R. Sridhar. Dual rail static CMOS architecture for wave pipelining. In *VLSI Design, 1996. Proceedings., Ninth International Conference on*, pages 335–336, Jan. 1996.

[Eis12] Holger Eisenreich. CoolBaseStations internal testchip documentation of Tomahawk2, TU Dresden. unpublished, 2012.

[EMH+07] H. Eisenreich, C. Mayr, S. Henker, M. Wickert, and R. Schüffny. A programmable clock generator HDL softcore. In *2007. MWSCAS 2007. 50th Midwest Symposium on Circuits and Systems*, pages 1–4, Aug. 2007.

[EMH+09] H. Eisenreich, C. Mayr, S. Henker, M. Wickert, and R. Schüffny. A novel ADPLL design using successive approximation frequency control. *Microelectron. J.*, 40:1613–1622, Nov. 2009.

[ES04] M. Elgebaly and M. Sachdev. Efficient adaptive voltage scaling system through on-chip critical path emulation. In *Low Power Electronics and Design, 2004. ISLPED '04. Proceedings of the 2004 International Symposium on*, pages 375 – 380, Aug. 2004.

[Fah05] AmrM. Fahim. *Clock Generators for SoC Processors.* Kluwer Academic Publishers, 2005.

[FKWG11] Xin Fan, M. Krstic, C. Wolf, and E. Grass. Gals design for on-chip ground bounce suppression. In *Asynchronous Circuits and Systems (ASYNC), 2011 17th IEEE International Symposium on*, pages 43 –52, Apr. 2011.

[Flo08] B.A. Floyd. Sub-integer frequency synthesis using phase-rotating frequency dividers. *Circuits and Systems I: Regular Papers, IEEE Transactions on*, 55(7):1823 –1833, Aug. 2008.

[GIK+09] M. Ghoneima, Y. Ismail, M.M. Khellah, J. Tschanz, and V. De. Serial-link bus: A low-power on-chip bus architecture. *Circuits and Systems I: Regular Papers, IEEE Transactions on*, 56(9):2020 –2032, Sep. 2009.

[GKGN09] Xiang Gao, E.A.M. Klumperink, P.F.J. Geraedts, and B. Nauta. Jitter analysis and a benchmarking figure-of-merit for phase-locked loops. *Circuits and Systems II: Express Briefs, IEEE Transactions on*, 56(2):117 –121, Feb. 2009.

[GNDD10] W. Grollitsch, R. Nonis, and N. Da Dalt. A 1.4ps rms-period-jitter TDC-less fractional-N digital PLL with digitally controlled ring oscillator in 65nm CMOS. In *Solid-State Circuits Conference Digest of Technical Papers (ISSCC), 2010 IEEE International*, pages 478 –479, Feb. 2010.

[Gra07] Helmut E. Graeb. *Analog Design Centering and Sizing*. Springer, 2007.

[Haa11] T. Haase. Entwurf eines ADPLL Taktgenerators für schnelle dynamische Frequenz-Skalierung. Studienarbeit, Technische Universität Dresden, 2011. Studienarbeit, Technische Universität Dresden, Betreuer: S. Höppner, R. Schüffny, in German.

[Hen03] J. Henkel. Closing the soc design gap. *Computer*, 36(9):119 – 121, Sep. 2003.

[HG12] Xuchu Hu and M.R. Guthaus. Distributed lc resonant clock grid synthesis. *Circuits and Systems I: Regular Papers, IEEE Transactions on*, 59(11):2749 –2760, Nov. 2012.

[HHN+10] H. Hamasaki, Y. Hoshi, A. Nakamura, A. Yamamoto, H. Kido, and S. Muramatsu. Soc for car navigation system with a 55.3gops image recognition engine. In *Design Automation Conference (ASP-DAC), 2010 15th Asia and South Pacific*, pages 464 –465, Jan. 2010.

[HL09] Chao-Ching Hung and Shen-Iuan Liu. A leakage-compensated PLL in 65-nm CMOS technology. *Circuits and Systems II: Express Briefs, IEEE Transactions on*, 56(7):525 –529, Jul. 2009.

[HMY10] P.-H. Hsieh, J. Maxey, and C.-K. K. Yang. A phase-selecting digital phase-locked loop with bandwidth tracking in 65-nm CMOS technology. *Solid-State Circuits, IEEE Journal of*, 45(4):781 –792, Apr. 2010.

[Hof12] K. Hofmann. Network-on-chip: Challenges for the interconnect and i/o architecture. In *High Performance Computing and Simulation (HPCS), 2012 International Conference on*, pages 252 –253, Jul. 2012.

[HOH+08] Ron Ho, T. Ono, R.D. Hopkins, A. Chow, J. Schauer, F.Y. Liu, and R. Drost. High speed and low energy capacitively driven on-chip wires. *Solid-State Circuits, IEEE Journal of*, 43(1):52 –60, Jan. 2008.

[Höp08] S. Höppner. Development of components for a fractional-N-PLL in CMOS-technology. Diplomarbeit, Technische Universität Dresden, 2008. Diploma Thesis, Technische Universität Dresden, Betreuer: S. Henker, R. Schüffny.

[HWB04] Frank Herzel, Wolfgang Winkler, and Johannes Borngräber. Jitter and phase noise in oscillators and phase-locked loops. In *Proc. SPIE Second International Symposium on Fluctuations and Noise*, 2004.

[INS+12] Y. Ikenaga, M. Nomura, S. Suenaga, H. Sonohara, Y. Horikoshi, T. Saito, Y. Ohdaira, Y. Nishio, T. Iwashita, M. Satou, K. Nishida, K. Nose, K. Noguchi, Y. Hayashi, and M. Mizuno. A 27% active-power-reduced 40-nm CMOS multimedia SoC with adaptive voltage scaling using distributed universal delay lines. *Solid-State Circuits, IEEE Journal of*, 47(4):832 -840, Apr. 2012.

[ITR11b] ITRS 2011 edition, international technology roadmap for semiconductors, design, 2011.

[ITR11c] ITRS 2011 edition, international technology roadmap for semiconductors, executive summary, 2011.

[JAPR12] T. Jungeblut, J. Ax, M. Porrmann, and U. Rückert. A TCMS-based architecture for GALS NoCs. In *Circuits and Systems (ISCAS), 2012 IEEE International Symposium on*, pages 2721 -2724, May 2012.

[JED07] JEDEC SOLID STATE TECHNOLOGY ASSOCIATION. *JESD208 SPECIALITY DDR2-1066 SDRAM*, November 2007.

[JED09] JEDEC SOLID STATE TECHNOLOGY ASSOCIATION. *JESD79-2F DDR2 SDRAM SPECIFICATION*, November 2009.

[JED10] JEDEC SOLID STATE TECHNOLOGY ASSOCIATION. *JESD79-3E DDR3 SDRAM SPECIFICATION*, July 2010.

[Jip08] R. Jipa. Dedicated solution for local clock programing in GALS designs. In *Semiconductor Conference, 2008. CAS 2008. International*, volume 2, pages 393 -396, Oct. 2008.

[JSC+12] Dongsuk Jeon, Mingoo Seok, C. Chakrabarti, D. Blaauw, and D. Sylvester. A super-pipelined energy efficient subthreshold 240 ms/s fft core in 65 nm cmos. *Solid-State Circuits, IEEE Journal of*, 47(1):23 -34, Jan. 2012.

[KFA+07] Michael Keating, David Flynn, Robert Aitken, Alan Gibbons, and Kaijian Shi. *Low Power Methodology Manual For System-on-Chip Design.* Springer, 2007.

[KFG+11] M. Krstic, X. Fan, E. Grass, C. Heer, B. Sanders, L. Benini, M.R. Kakoee, A. Strano, and D. Bertozzi. Moonrake chip - gals demonstrator in 40 nm CMOS technology. In *System on Chip (SoC), 2011 International Symposium on*, pages 9 -13, 31 2011-nov. 2 2011.

[KGWB08] Wonyoung Kim, M.S. Gupta, Gu-Yeon Wei, and D. Brooks. System level analysis of fast, per-core DVFS using on-chip switching regulators. In *High Performance Computer Architecture, 2008. HPCA 2008. IEEE 14th International Symposium on*, pages 123 –134, Feb. 2008.

[KKK+06] J. H. Kim, Y. H. Kwak, M. Kim, S. W. Kim, and C. Kim. A 120 MHz 1.8 GHz CMOS DLL based clock generator for dynamic frequency scaling. *IEEE Journal of Solid-State Circuits*, 41(9):2077–2082, Sep. 2006.

[KMN+09] N. Kurd, P. Mosalikanti, M. Neidengard, J. Douglas, and R. Kumar. Next generation intel core micro-architecture (nehalem) clocking. *Solid-State Circuits, IEEE Journal of*, 44(4):1121 –1129, Apr. 2009.

[KSK+09] Deok-Soo Kim, Heesoo Song, Taeho Kim, Suhwan Kim, and Deog-Kyoon Jeong. A 1.35GHz all-digital fractional-n PLL with adaptive loop gain controller and fractional divider. In *Solid-State Circuits Conference, 2009. A-SSCC 2009. IEEE Asian*, pages 161 –164, Nov. 2009.

[Kun05] Ken Kundert. Predicting the phase noise and jitter of PLL-based frequency synthesizers. *www.designers-guide.org*, 2005.

[KZS11] T. Kolpe, A. Zhai, and S.S. Sapatnekar. Enabling improved power management in multicore processors through clustered dvfs. In *Design, Automation Test in Europe Conference Exhibition (DATE), 2011*, pages 1 –6, Mar. 2011.

[LCL09] Wei-Ming Lin, Chao-Chyun Chen, and Shen-Iuan Liu. An all-digital clock generator for dynamic frequency scaling. In *VLSI Design, Automation and Test, 2009. VLSI-DAT '09. International Symposium on*, pages 251 –254, Apr. 2009.

[Lee02] David C. Lee. Analysis of jitter in phase-locked loops. *IEEE Transactions on Circuits and Systems*, 49(11):704 – 711, 2002.

[Lev04] L. Lev. Mind the design gap. *IEE Review*, 50(10):37, Oct. 2004.

[LJB+13] S. Lütkemeier, T. Jungeblut, H. K. O. Berge, S. Aunet, M. Porrmann, and U. Rückert. A 65 nm 32 b subthreshold processor with 9T multi-vt SRAM and adaptive supply voltage control. *Solid-State Circuits, IEEE Journal of*, 48(1):8 –19, Jan. 2013.

[LJK+05] Kwang-Jin Lee, Seung-Hun Jung, Yun-Jeong Kim, Chul Kim, Suki Kim, Uk-Rae Cho, Choong-Guen Kwak, and Hyun-Geun Byun. A digitally controlled oscillator for low jitter all digital phase locked loops. In *Asian Solid-State Circuits Conference, 2005*, pages 365 –368, Nov. 2005.

[LOK+12] Y.W. Li, C. Ornelas, Hyung Seok Kim, H. Lakdawala, A. Ravi, and K. Soumyanath. A reconfigurable distributed all-digital clock generator core with SSC and skew correction in 22nm high-k tri-gate LP CMOS. In *Solid-State Circuits Conference Digest of Technical Papers (ISSCC), 2012 IEEE International*, pages 70 –72, Feb. 2012.

[Lu93] S.L. Lu. Improved design of CMOS multiple-input Muller-C-elements. *Electronics Letters*, 29(19):1680 –1682, Sep. 1993.

[LWB+08] T. Limberg, M. Winter, M. Bimberg, R. Klemm, E. Matus, M.B.S. Tavares, G. Fettweis, H. Ahlendorf, and P. Robelly. A fully programmable 40 GOPS SDR single chip baseband for LTE/WiMAX terminals. In *Solid-State Circuits Conference, 2008. ESSCIRC 2008. 34th European*, pages 466 –469, Sep. 2008.

[MB10] Dongsheng Ma and R. Bondade. Enabling power-efficient DVFS operations on silicon. *Circuits and Systems Magazine, IEEE*, 10(1):14 –30, 2010.

[MGS08] T. Massier, H. Graeb, and U. Schlichtmann. The sizing rules method for CMOS and bipolar analog integrated circuit synthesis. *Computer-Aided Design of Integrated Circuits and Systems, IEEE Transactions on*, 27(12):2209–2222, Dec. 2008.

[MKO+12] H. Miyazaki, Y. Kusano, H. Okano, T. Nakada, K. Seki, T. Shimizu, N. Shinjo, F. Shoji, A. Uno, and M. Kurokawa. K computer: 8.162 petaflops massively parallel scalar supercomputer built with over 548k cores. In *Solid-State Circuits Conference Digest of Technical Papers (ISSCC), 2012 IEEE International*, pages 192 –194, Feb. 2012.

[MNS03] M. Maymandi-Nejad and M. Sachdev. A digitally programmable delay element: design and analysis. *Very Large Scale Integration (VLSI) Systems, IEEE Transactions on*, 11(5):871 –878, Oct. 2003.

[MPPdG04] M. Meijer, F. Pessolano, and J. Pineda de Gyvez. Technology exploration for adaptive power and frequency scaling in 90nm CMOS. In *Low Power Electronics and Design, 2004. ISLPED '04. Proceedings of the 2004 International Symposium on*, pages 14 – 19, Aug. 2004.

[MR09] John A. McNeill and David Ricketts. *The Designer's Guide to Jitter in Ring Oscillators*. Springer, 2009.

[MSK+05] E. Mensink, D. Schinkel, E. Klumperink, E. van Tuijl, and B. Nauta. Optimally-placed twists in global on-chip differential interconnects. In *Solid-State Circuits Conference, 2005. ESSCIRC 2005. Proceedings of the 31st European*, pages 475 – 478, Sep. 2005.

[MSK+10] E. Mensink, D. Schinkel, E.A.M. Klumperink, E. van Tuijl, and B. Nauta. Power efficient gigabit communication over capacitively driven RC-limited on-chip interconnects. *Solid-State Circuits, IEEE Journal of*, 45(2):447 – 457, Feb. 2010.

[MTC+00] S.W. Moore, G.S. Taylor, P.A. Cunningham, R.D. Mullins, and P. Robinson. Self calibrating clocks for globally asynchronous locally synchronous systems. In *Computer Design, 2000. Proceedings. 2000 International Conference on*, pages 73 –78, 2000.

[MX00] H. Mair and Liming Xiu. An architecture of high-performance frequency and phase synthesis. *Solid-State Circuits, IEEE Journal of*, 35(6):835–846, Jun. 2000.

[NS10] A. Narasimhan and R. Sridhar. Variability aware low-power delay optimal buffer insertion for global interconnects. *Circuits and Systems I: Regular Papers, IEEE Transactions on*, 57(12):3055 –3063, Dec. 2010.

[ON04] T. Olsson and P. Nilsson. A digitally controlled PLL for SoC applications. *Solid-State Circuits, IEEE Journal of*, 39(5):751 – 760, May 2004.

[PK13] M.-J. Park and J. Kim. Pseudo-linear analysis of bang-bang controlled timing circuits. *Circuits and Systems I: Regular Papers, IEEE Transactions on*, PP(99):1, 2013.

[PKGN12] X. Pu, A. Kumar, S. Goldman, and K. Nagaraj. Low-noise low-spur architecture for a fully integrated analog PLL working from a low-frequency reference. *Circuits and Systems II: Express Briefs, IEEE Transactions on*, PP(99):1 –5, 2012.

[PKPF09] JunYoung Park, J. Kang, Sunghyun Park, and M.P. Flynn. A 9-Gbit/s serial transceiver for on-chip global signaling over lossy transmission lines. *Circuits and Systems I: Regular Papers, IEEE Transactions on*, 56(8):1807 –1817, Aug. 2009.

[PMP09] S. Pellerano, P. Madoglio, and Y. Palaskas. A 4.75-GHz fractional frequency divider-by-1.25 with TDC-based all-digital spur calibration in 45-nm CMOS. *Solid-State Circuits, IEEE Journal of*, 44(12):3422–3433, Dec. 2009.

[QPP94] J. Qian, S. Pullela, and L. Pillage. Modeling the effective capacitance for the RC interconnect of CMOS gates. *Computer-Aided Design of Integrated Circuits and Systems, IEEE Transactions on*, 13(12):1526–1535, Dec. 1994.

[Ram07] Ulrich Ramacher. Software-defined radio prospects for multistandard mobile phones. *Computer*, 40(10):62–69, 2007.

[RBB+11] R.J. Riedlinger, R. Bhatia, L. Biro, B. Bowhill, E. Fetzer, P. Gronowski, and T. Grutkowski. A 32nm 3.1 billion transistor 12-wide-issue itanium processor for mission-critical servers. In *Solid-State Circuits Conference Digest of Technical Papers (ISSCC), 2011 IEEE International*, pages 84 –86, Feb. 2011.

[RRH+11] U. Ramacher, W. Raab, U. Hachmann, D. Langen, J. Berthold, R. Kramer, A. Schackow, C. Grassmann, M. Sauermann, P. Szreder, F. Capar, G. Obradovic, W. Xu, N. Bruls, Kang Lee, Eugene Weber, Ray Kuhn, and John Harrington. Architecture and implementation of a software-defined radio baseband processor. In *Circuits and Systems (ISCAS), 2011 IEEE International Symposium on*, pages 2193 –2196, May 2011.

[RTE+08] A. Rylyakov, J. Tierno, G. English, M. Sperling, and D. Friedman. A wide tuning range (1 GHz-to-15 GHz) fractional-N all-digital PLL in 45nm SOI. In *Custom Integrated Circuits Conference, 2008. CICC 2008. IEEE*, pages 431 –434, Sep. 2008.

[SAA06] K. Sundaresan, P.E. Allen, and F. Ayazi. Process and temperature compensation in a 7-MHz CMOS clock oscillator. *Solid-State Circuits, IEEE Journal of*, 41(2):433 – 442, Feb. 2006.

[SAI+13] V. S. Sathe, S. Arekapudi, A. Ishii, C. Ouyang, M. C. Papaefthymiou, and S. Naffziger. Resonant-clock design for a power-efficient, high-volume x86-64 microprocessor. *Solid-State Circuits, IEEE Journal of*, 48(1):140 –149, Jan. 2013.

[SB07] Robert Bogdan Staszewski and Poras T. Balsara. All-digital PLL with ultra fast settling. *Circuits and Systems II: Express Briefs, IEEE Transactions on*, 54(2):181 –185, Feb. 2007.

[SCJ+11] Seong-Young Seo, Jung-Hoon Chun, Young-Hyun Jun, Seok Kim, and Kee-Won Kwon. A digitally controlled oscillator with wide frequency range and low supply sensitivity. *Circuits and Systems II: Express Briefs, IEEE Transactions on*, 58(10):632 –636, Oct. 2011.

[SCL07] Duo Sheng, Ching-Che Chung, and Chen-Yi Lee. An ultra-low-power and portable digitally controlled oscillator for soc applications. *Circuits and Systems II: Express Briefs, IEEE Transactions on*, 54(11):954 –958, Nov. 2007.

[Sem97] Bronstein Semendyayev. *Handbook of Mathematics.* Springer, 1997.

[SH06] Kaijian Shi and D. Howard. Challenges in sleep transistor design and implementation in low-power designs. In *Design Automation Conference, 2006 43rd ACM/IEEE*, pages 113 –116, 0-0 2006.

[SHL+10] Jae-sun Seo, Ron Ho, J. Lexau, M. Dayringer, D. Sylvester, and D. Blaauw. High-bandwidth and low-energy on-chip signaling with adaptive preemphasis in 90nm cmos. In *Solid-State Circuits Conference Digest of Technical Papers (ISSCC), 2010 IEEE International*, pages 182 –183, Feb. 2010.

[SJJ+11] P. Salihundam, S. Jain, T. Jacob, S. Kumar, V. Erraguntla, Y. Hoskote, S. Vangal, G. Ruhl, and N. Borkar. A 2 Tb/s 6, *times*, 4 mesh network for a single-chip cloud computer with DVFS in 45 nm CMOS. *Solid-State Circuits, IEEE Journal of*, 46(4):757 –766, Apr. 2011.

[SKDM10] G. Shamanna, N. Kurd, J. Douglas, and M. Morrise. Scalable, sub-1W, sub-10ps clock skew, global clock distribution architecture for Intel Core i7/i5/i3 microprocessors. In *VLSI Circuits (VLSIC), 2010 IEEE Symposium on*, pages 83 –84, Jun. 2010.

[SKK08] Dongsuk Shin, Soo-Won Kim, and Chulwoo Kim. Wide frequency range duty cycle correction circuit for DDR interface. *IEICE Electronics Express*, 5(8):254–259, 2008.

[SLH+10] Gang-Neng Sung, Szu-Chia Liao, Jian-Ming Huang, Yu-Cheng Lu, and Chua-Chin Wang. All-digital frequency synthesizer using a flying adder. *Circuits and Systems II: Express Briefs, IEEE Transactions on*, 57(8):597 –601, Aug. 2010.

[SLM01] M. Saint-Laurent and G.P. Muyshondt. A digitally controlled oscillator constructed using adjustable resistors. In *Mixed-Signal Design, 2001. SSMSD. 2001 Southwest Symposium on*, pages 80 –82, 2001.

[SLP08] A.L. Sobczyk, A.W. Luczyk, and W.A. Pleskacz. Controllable local clock signal generator for deep submicron GALS architectures. In *Design and Diagnostics of Electronic Circuits and Systems, 2008. DDECS 2008. 11th IEEE Workshop on*, pages 1 –4, Apr. 2008.

[SM90] M. Soyuer and R.G. Meyer. Frequency limitations of a conventional phase-frequency detector. *Solid-State Circuits, IEEE Journal of*, 25(4):1019 – 1022, Aug. 1990.

[SMK+09] D. Schinkel, E. Mensink, E. Klumperink, E. van Tuijl, and B. Nauta. Low-power, high-speed transceivers for network-on-chip communication. *Very*

Large Scale Integration (VLSI) Systems, IEEE Transactions on, 17(1):12 –21, Jan. 2009.

[SN90] T. Sakurai and A.R. Newton. Alpha-power law mosfet model and its applications to CMOS inverter delay and other formulas. *Solid-State Circuits, IEEE Journal of*, 25(2):584 –594, Apr. 1990.

[SSS05] Keliu Shu and Edgar Sinchez-Sinencio. *CMOS PLL Synthesizers: Analysis and Design*. Springer, 2005.

[TCM+09] D.N. Truong, W.H. Cheng, T. Mohsenin, Zhiyi Yu, A.T. Jacobson, G. Landge, M.J. Meeuwsen, C. Watnik, A.T. Tran, Zhibin Xiao, E.W. Work, J.W. Webb, P.V. Mejia, and B.M. Baas. A 167-processor computational platform in 65 nm CMOS. *Solid-State Circuits, IEEE Journal of*, 44(4):1130 –1144, Apr. 2009.

[TLC+10] Chao-Fang Tsai, Wan-Jing Li, Peng-Yu Chen, Ying-Zu Lin, and Soon-Jyh Chang. On-chip reference oscillators with process, supply voltage and temperature compensation. In *Next-Generation Electronics (ISNE), 2010 International Symposium on*, pages 108 –111, Nov. 2010.

[TRF08] J. A. Tierno, A. V. Rylyakov, and D. J. Friedman. A wide power supply range, wide tuning range, all static CMOS all digital PLL in 65 nm SOI. *IEEE Journal of Solid-State Circuits*, 43(1):42–51, Jan. 2008.

[Uye01] John P. Uyemura. *CMOS Logic Circuit Design*. Kluwer Academic Publishers, 2001.

[vdBKVN02] R.C.H. van de Beek, E.A.M. Klumperink, C.S. Vaucher, and B. Nauta. Low-jitter clock multiplication: a comparison between PLLs and DLLs. *Circuits and Systems II: Analog and Digital Signal Processing, IEEE Transactions on*, 49(8):555 – 566, Aug. 2002.

[vdTKvR03] Johan van der Tang, Dieter Kasperovitz, and Arthur van Roermund. *High-Frequency Oscillator Design for Integrated Transceivers*. Kluwer Academic Publishers, 2003.

[VFL+00] C. S. Vaucher, I. Ferencic, M. Locher, S. Sedvallson, U. Voegeli, and Z. Wang. A family of low-power truly modular programmable dividers in standard 0.35-μm CMOS technology. *IEEE Journal of Solid-State Circuits*, 35(7):1039–1045, Jul. 2000.

[VHR+08] S.R. Vangal, J. Howard, G. Ruhl, S. Dighe, H. Wilson, J. Tschanz, D. Finan, A. Singh, T. Jacob, S. Jain, V. Erraguntla, C. Roberts, Y. Hoskote, N. Borkar, and S. Borkar. An 80-tile sub-100-W teraflops processor in

65-nm CMOS. *Solid-State Circuits, IEEE Journal of*, 43(1):29 -41, Jan. 2008.

[Wag09] T. Wagner. Entwurf einer ADPLL Schaltung in 45 nm CMOS Technologie. Diplomarbeit, Technische Universität Dresden, 2009. Diploma Thesis, Technische Universität Dresden, Betreuer: S. Höppner, S. Henker, R. Schüffny, in German.

[Wal10] D. Walter. Serielle Punkt-zu-Punkt Verbindungen für MPSoC NoCs. Diplomarbeit, Technische Universität Dresden, 2010. Diploma Thesis, Technische Universität Dresden, Betreuer: H. Eisenreich, S. Höppner, R. Schüffny, in German.

[War11] J. Warnock. Circuit design challenges at the 14nm technology node. In *Design Automation Conference (DAC), 2011 48th ACM/EDAC/IEEE*, pages 464 -467, Jun. 2011.

[Win10] Markus Winter. CoolBaseStations internal testchip documentation of Tommy and Atlas, TU Dresden. unpublished, 2010.

[Win11] M. Winter. *Unterstützung und Organisation von Quality-of-Service Techniken in Kommunikationsnetzwerken auf einem Chip (Network-on-Chip)*. Dissertation, 2011. Technische Universität Dresden, in German.

[WLL+09] Shien-Yang Wu, J.J. Liaw, C.Y. Lin, M.C. Chiang, C.K. Yang, J.Y. Cheng, M.H. Tsai, M.Y. Liu, P.H. Wu, C.H. Chang, L.C. Hu, C.I. Lin, H.F. Chen, S.Y. Chang, S.H. Wang, P.Y. Tong, Y.L. Hsieh, K.H. Pan, C.H. Hsieh, C.H. Chen, C.H. Yao, C.C. Chen, T.L. Lee, C.W. Chang, H.J. Lin, S.C. Chen, J.H. Shieh, S.M. Jang, K.S. Chen, Y. Ku, Y.C. See, and W.J. Lo. A highly manufacturable 28nm CMOS low power platform technology with fully functional 64Mb SRAM using dual/tripe gate oxide process. In *VLSI Technology, 2009 Symposium on*, pages 210 -211, Jun. 2009.

[WPG10] M. Winter, S. Prusseit, and P.F. Gerhard. Hierarchical routing architectures in clustered 2D-mesh networks-on-chip. In *SoC Design Conference (ISOCC), 2010 International*, pages 388 -391, Nov. 2010.

[WSWW10] Chia-Tsun Wu, Wen-Chung Shen, Wei Wang, and An-Yeu Wu. A two-cycle lock-in time ADPLL design based on a frequency estimation algorithm. *Circuits and Systems II: Express Briefs, IEEE Transactions on*, 57(6):430 -434, Jun. 2010.

[WWWW05] Chia-Tsun Wu, Wei Wang, I-Chyn Wey, and An-Yeu Wu. A scalable DCO design for portable ADPLL designs. In *Circuits and Systems, 2005. ISCAS*

2005. IEEE International Symposium on, pages 5449–5452 Vol. 6, May 2005.

[WYZF07] Liang Wang, Suge Yue, Yuanfu Zhao, and Long Fan. An SEU-tolerant programmable frequency divider. In *Quality Electronic Design, 2007. ISQED '07. 8th International Symposium on*, pages 899 –904, Mar. 2007.

[WZQW09] Shengyang Wang, Jiafeng Zhu, Zhihua Qu, and Jianhui Wu. Power efficient multimodulus programmable frequency divider with half-integer division ratio step size. In *Electronics, Circuits, and Systems, 2009. ICECS 2009. 16th IEEE International Conference on*, pages 739 –742, Dec. 2009.

[Xiu07] Liming Xiu. A flying-adder on-chip frequency generator for complex SoC environment. *Circuits and Systems II: Express Briefs, IEEE Transactions on*, 54(12):1067 –1071, Dec. 2007.

[XLL12] Liming Xiu, Kun-Ho Lin, and Ming Lin. The impact of input-mismatch on flying-adder direct period synthesizer output jitter. *Circuits and Systems I: Regular Papers, IEEE Transactions on*, 59(9):1942 –1951, Sep. 2012.

[XY03] Liming Xiu and Zhihong You. A new frequency synthesis method based on "flying-adder" architecture. *Circuits and Systems II: Analog and Digital Signal Processing, IEEE Transactions on*, 50(3):130 – 134, Mar. 2003.

[XY05] Liming Xiu and Zhihong You. A "flying-adder" frequency synthesis architecture of reducing VCO stages. *Very Large Scale Integration (VLSI) Systems, IEEE Transactions on*, 13(2):201 –210, Feb. 2005.

[YB09] Zhiyi Yu and B.M. Baas. High performance, energy efficiency, and scalability with GALS chip multiprocessors. *Very Large Scale Integration (VLSI) Systems, IEEE Transactions on*, 17(1):66 –79, Jan. 2009.

[YCYL12] Chien-Ying Yu, Ching-Che Chung, Chia-Jung Yu, and Chen-Yi Lee. A low-power DCO using interlaced hysteresis delay cells. *Circuits and Systems II: Express Briefs, IEEE Transactions on*, 59(10):673 –677, Oct. 2012.

[YIE+11] W. Yin, R. Inti, A. Elshazly, B. Young, and P. K. Hanumolu. A 0.7-to-3.5 GHz 0.6-to-2.8 mW highly digital phase-locked loop with bandwidth tracking. *Solid-State Circuits, IEEE Journal of*, PP(99):1, 2011.

[YYG08] Yi Yang, LiQiong Yang, and Zhuo Gao. A PVT tolerant sub-mA PLL in 65nm CMOS process. In *Electronics, Circuits and Systems, 2008. ICECS 2008. 15th IEEE International Conference on*, pages 998 –1001, Sep. 2008.

[YYL11] C.-Y. Yu, J.-Y. Yu, and C.-Y. Lee. A low voltage all-digital on-chip oscillator using relative reference modeling. *Very Large Scale Integration (VLSI) Systems, IEEE Transactions on*, PP(99):1 -6, 2011.

[ZA11] Xuan Zhang and A.B. Apsel. A low-power, process-and- temperature- compensated ring oscillator with addition-based current source. *Circuits and Systems I: Regular Papers, IEEE Transactions on*, 58(5):868 -878, May 2011.

[ZAJ+11] E. Zianbetov, F. Anceau, M. Javidan, D. Galayko, E. Colinet, and J. Juillard. A digitally controlled oscillator in a 65-nm CMOS process for SoC clock generation. In *Circuits and Systems (ISCAS), 2011 IEEE International Symposium on*, pages 2845 -2848, May 2011.

[Zhe11] K. Zheng. Hardware Performance Monitor Design in Sub-100nm-Technologien. Diplomarbeit, Technische Universität Dresden, 2011. Studienarbeit, Technische Universität Dresden, Betreuer: H. Eisenreich, S. Höppner, R. Schüffny, in German.

[ZK08] Jun Zhao and Yong-Bin Kim. A 12-bit digitally controlled oscillator with low power consumption. In *Circuits and Systems, 2008. MWSCAS 2008. 51st Midwest Symposium on*, pages 370 -373, Aug. 2008.

[ZTL+09] M. Zanuso, D. Tasca, S. Levantino, A. Donadel, C. Samori, and A. L. Lacaita. Noise analysis and minimization in bang-bang digital PLLs. *Circuits and Systems II: Express Briefs, IEEE Transactions on*, 56(11):835-839, Nov. 2009.

About the author

Sebastian Höppner (* 19.06.1982) received the Dipl.-Ing. (M.Sc.) in Electrical Engineering from Technische Universität Dresden, Germany in 2008. In 2007 he worked at National Taiwan University Taipei, Taiwan as engineering intern on RF-IC design. From 2007 to 2008 he was with Gärtner Electronic Design GmbH (ELMOS AG) in Frankfurt (Oder), Germany. Since 2008 he has been a research associate and technical project manager with the Chair of Highly-Parallel VLSI-Systems and Neuromorphic Circuits at Technische Universität Dresden. His research interests include circuit design for clocking and data transmission in low-power systems-on-chip and design methodology for custom circuits in advanced CMOS technology nodes.